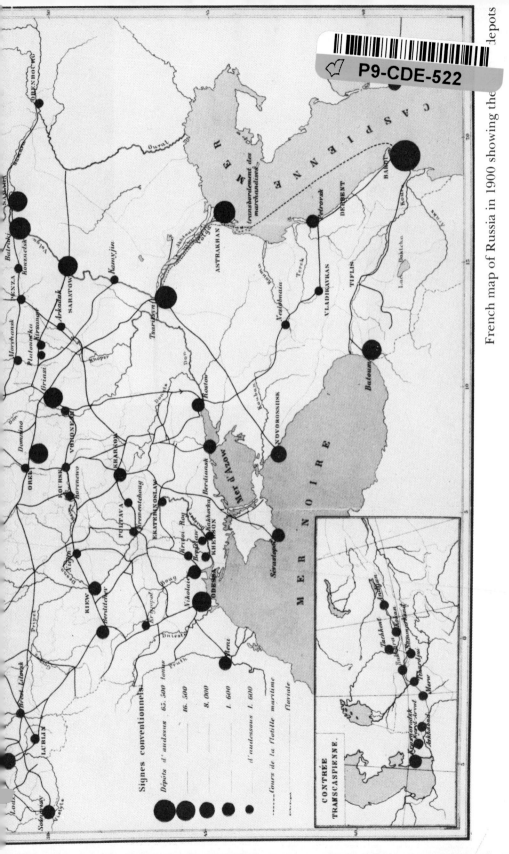

French map of Russia in 1900 showing the depots

DATE DUE

THE RUSSIAN
ROCKEFELLERS

The Saga of the Nobel Family and the Russian Oil Industry

Robert W. Tolf

Hoover Institution Press
Stanford University
Stanford, California

Hoover Institution Publication 158

International Standard Book Number 0-8179-6581-5
Library of Congress Card Number 76-284
Printed in the United States of America
© 1976 by the Board of Trustees of the
  Leland Stanford Junior University
All rights reserved

Contents

To Nancy

Preface

Robert Tolf has written a book that is unique in its field, a book that will be a blessing to scholars and fascinating to the general reader. We all know who Alfred Nobel is, largely because of the Nobel prizes for work in furthering international peace or for excellence in physics, chemistry, literature, or economics. A number of people know, too, of Nobel's connection with the invention of dynamite. But who knows the other members of the Nobel family, their remarkable contributions to invention and transportation, their building of the world's greatest oil industry, their aid to speedy growth of industrialization and modernization of Russia under the last tsar? The books that have hitherto been published deal principally with the prizes Alfred Nobel's foundation has awarded and the men and women who have won them. But Dr. Tolf has done the first book which tells the whole saga of the Nobel family and the amazing work they accomplished in developing the oil industry of Russia—inventing the most important devices in the refining, the storing, the shipping by pipe line and oil tanker—in short, the very foundations of our modern industry which we only became aware of when the Arab consortium multiplied the price of crude oil fivefold.

In *The Russian Rockefellers,* using Swedish, Norwegian, Russian, German, English, American, and French sources, the author rescues a great, constructive, venturous, and inventive family from the darkness into which the Soviet leaders, in their manipulation of history, have sought to plunge their works and days. The members of this family whose story is told here made Baku into the world's greatest oil producing center; built decent houses for their workers and created decent working conditions; showed honor, integrity, and benevolence in trade. The head of the family, Immanuel Nobel, built a munitions factory in old Russia that specialized in such things as undersea mines. When he took young Alfred, as well as his other sons, into the explosive business, Alfred proved a remarkable inventor, too, blowing up an entire factory when he was working on nitroglycerine. Alfred then slowed up this dangerous explosive with wood shavings, sawdust, and finally with earth composed of the silicon remains of diatoms, thus producing dynamite. But, being of a pacifist inclination, Alfred left his fortune to his foundation chiefly to give a prize each year to the greatest contributor to international peace.

Before the beginning of World War I, Russia was producing more than half the world's oil and the Nobel family one-third of Russia's crude and 40 percent of its refined petroleum. But before the war's end, one Nobel was fleeing Russia disguised as a peasant, and two of his brothers were in the cellars of the Cheka, while their great industrial empire was in ruins. Until now, the Nobel story has been largely shrouded in darkness because Lenin and Stalin in seizing power have kept it dark, as they have kept the mass of Russians in darkness. They have renamed the factories, harbors, dwellings and rest homes after Bolshevik leaders and done their best to make the Nobels into unpersons and their works into apparent achievements of the Soviet regime. But in the long run, history will not be so cheated. Dr. Tolf has made a remarkable contribution to the proof of that fact.

Bertram Wolfe
Distinguished Professor Emeritus of Russian History
University of California

Senior Research Fellow
Hoover Institution
Stanford University

Introduction

Most of the civilized world has heard of Alfred Nobel and the prizes given annually in his name, but he was only one member of history's most inventive family and the story has never been told of his father Immanuel, his brothers Ludwig and Robert, his nephew Emanuel, and their own individual entrepreneurial, technological, and financial achievements in the armaments, petroleum, chemical, and transportation industries.

Ludwig Nobel, after a highly successful career as a St. Petersburg manufacturer, literally created the Russian oil industry which in turn fueled the tremendous economic expansion of prerevolutionary Russia. He designed the world's first oil tanker and had a dozen in regular service before other nations followed his lead; he installed Europe's first pipelines and put the first tankcars on its rails; he built the world's first full-scale continuous distillation refinery and developed oil burners to utilize more efficiently the black gold pouring from the world's first gushers; he forged a gigantic infrastructure on water and land, overcoming native inertia and a stubborn opposition, building a network of storage depots and tank farms, harbors, freightyards, and marketing outlets from one end of the vast Russian empire to the other and then across the European continent and into the British Isles.

Ludwig's father before him pioneered development of underwater mines, designed some of the first steam engines to power Russian ships, and installed the first central heating systems to warm Russian homes. Ludwig's son after him launched the world's first diesel-driven tugs and tankers while bargaining with the Rothschilds, struggling against Royal Dutch–Shell, and bartering with Standard Oil in Europe's second Thirty Years War, a petroleum war for control of world markets.

During their eighty years in Russia the Nobels and their band of Scandinavian expatriates were industrial miracle makers in a primitive land.

They built armament factories in the wilds of northern Russia, expanded a small St. Petersburg machine shop and foundry into one of the largest enterprises in the country, developed and marketed the Nobel wheel—the Michelin of its time—devised new tools and procedures for assembly-line production, exhibiting in all their undertakings an active and continuing concern for their workers' welfare—as unique a concern in Russia at the time as it was in the rest of the world. Nobels pioneered decent housing, medical care, technical training for the workers, and elementary education for their children. They abolished child labor and shortened work hours, while establishing workers' savings banks and a system of regular wage payments without resort to the standard and iniquitous system of fines for real or imagined transgressions. Their fifty thousand workers felt a special loyalty and pride in being identified as "Nobelites."

The Nobels worked their greatest wonders in Baku, that ancient crossroad of civilizations which Norwegian writer Knut Hamsun found "much too Persian to be European and much too European to be Persian." Baku was part pure *Arabian Nights* and part Byzantine boom town of baksheesh and brutality. The feuds of Armenians and Tatars, the barbaric rampages of the most degenerate forces of reaction and repression, brought death and destruction. Blood ran in the streets, mingling with the sweet smell of oil and the sooty smoke belching from the dozens of refineries ringing the city in what became known as Black Town.

It was commonly said in Russia that among the two hundred oil barons of Baku only ten were honest: Nobel, an Armenian, and eight Moslems. Nobel was also the king, the dominant force, controlling transport across the Caspian and up the mighty Volga and sharing with the Rothschilds control of the Transcaucasian route to the Black Sea port of Batum. The Nobel Brothers Petroleum Company was the innovator and the leader: the others followed, first in the empire and then in Europe and in Asia. The family's achievements brought it to the pinnacle of power and fame. Typical was the judgment of the editor of London's *Petroleum World:* "The history of the Nobel family, written fully, would have an interest for all who admire the triumphs of genius in trade, associated with perfect honour and integrity."

That history has not been written because in Baku other forces were at work, a subculture that eventually emerged to seize power and to cast the Nobels and all their Russian achievements and contributions into a limbo of the lost.

Baku was where Stalin received what he termed his "revolutionary baptism in combat," where he "learned what it meant to lead great masses of workers" in that "storm of the deepest conflicts between workers and oil industrialists." Stalin, Kalinin, Beria, Leonid Krassin, Lev Kamenev, Stepan Shaumyan, and countless others received their basic training in the city which became known as "the revolutionary hotbed on the Caspian." The oil workers of Baku sparked the country's first general strike, long before the revolution of 1905. Long after that revolution, when the rest of Russia was crushed once again into apathetic acceptance of autocracy, orthodoxy, and nationalism, those workers were still using the strike to gain their ends, earning Lenin's praise as "our last Mohicans of the political mass strike."

At the turn of the century Russia was supplying more than half the world's oil, but the troubles of 1903–1906 marked the end of that leadership and spelled defeat for Nobel, Rothschild, and the other Russian producers in the Thirty Years War. Decreased production made possible a new era of international cooperation among the oil companies with attendant price-fixing and market-sharing agreements. It also meant record profits while Emanuel Nobel, a director of the State Bank with excellent connections at court and in several ministries, led the official and private rehabilitation of an industry decimated by the ravages of revolution. He introduced new American rotary drilling techniques, developed new oil fields, constructed new chemical plants, and negotiated the Russian rights for the diesel engine. In the years before the First World War his engineers built and installed dozens of diesels in power plants, factories, and pumping stations along the newly completed Transcaucasian pipeline and in tugs and tankers working the Caspian-Volga, Black sea, and Baltic routes. As his grandfather had modernized the Russian navy with new engines in the 1850s so Emanuel developed new engines for the navy fifty years later, establishing a factory to build diesel submarines.

By 1916 Nobel owned, controlled, or had substantial interest in companies producing a third of all Russian crude oil, 40 percent of all the refined, and supplying almost two-thirds of domestic consumption. There were more than four hundred tank farms and depots flying the Nobel banner and the company commanded the largest private fleet in the world.

Two years later Emanuel was fleeing the country disguised as a peasant, his two brothers were in a Cheka prison, the Nobel empire was a shambles, the ships halted, the refinery fires banked, the hundreds of wells filling with water, the factories in St. Petersburg closed down.

For the next decade in exile, first in Paris and then in Stockholm, the Nobels fought to recover what had been seized by the Bolsheviks. But as the Soviets with Lenin's New Economic Policy and with massive injections of Western aid worked the miracle of economic revival, the Nobels gradually realized the futility of their position. It was a time when promoters traded pharmaceuticals for priceless art at bargain basement prices, when wheeler-dealers squirmed to secure exclusive exploitation rights, when bankers and industrialists circumvented their own governments' regulations in the rush to gain concessions. And the Soviets, led by the same Krassin who had received his revolutionary baptism in Baku, outmaneuvered their opponents. As the European countries lined up to fill their tanks with Russian oil, the fate of the Nobels was sealed; the Russian Rockefellers became simply the Swedish Nobels, destined to fade away in the shadow of Alfred.

Emanuel died in 1932 and the company was finally dissolved in 1959, its eightieth year. By that time the Soviets had obliterated all traces of Nobel and their works were forgotten. As the prerevolutionary slate was wiped clean, the history of an entire industry was lost and with it the record of those who had led that industry, those who had done so much for the economic development of Russia. There is no room in Soviet texts for Nobel nor is there credit for industrial innovation before 1917. The standard Soviet work, *Development and Exploitation of Oil and Gas Fields*, does not mention the name Nobel or any of the Nobel contributions, but dismisses the entire prerevolutionary period as one in which there was little technological advance. Similarly, and for the mass audience, the September 1972 issue of *Soviet Life* devoted to Transcaucasia ignores the subject by declaring that the area was "largely agricultural before the Revolution." The Communists have their own heroes and there were enough who fought and struggled in the oil-rich lands of Baku to fill their own hall of fame.

Revolution, nationalization, and war destroyed the records, and the double-think historians spun new versions to fit the party mold. But some contemporary accounts and documents did survive, and several of the exiled veterans put to paper their memories of the years in Russia. The Landsarkiv in Lund, Sweden, has a collection of manuscript material pertaining primarily to the activities of Immanuel and Robert, and I am grateful to the staff there for facilitating my use of those papers. I am especially indebted to Baron Stig Ramel and the Nobel Foundation in Stockholm for the privilege of studying the pertinent papers in the Alfred Nobel Collection. The always helpful staff in the Reading Room of the British Museum made my weeks of research there as pleasant as they were productive, and I also received from the staffs of the libraries of Miami and Boca Raton, the University of Miami, and Florida Atlantic University, the kind of support that eases and expedites the scholar's burdens. Mr. John Munday, Director of Naval Weapons and Antiquities at the British National Maritime Museum, provided helpful information on Nobel mines, and Dr. Philip K. Lundeberg, Curator of Naval History at the Smithsonian Institution, kindly made available the results of his own extensive research in the sometimes arcane subject of underwater mine warfare. Dr. Nils Oleinikoff, grandson of Ludwig Nobel, and Boris Hagelin, son of Karl Wilhelm who was Emanuel Nobél's closest associate, generously provided photographs and information from their own Russian experiences. Sallie Jaggar Hayes provided enthusiastic and expert editing. The responsibility for judgments and interpretations in the final product is, of course, as much mine as is my deep sense of appreciation.

Finally, a word for my wife who has put up rather patiently these past three years with my wanderings in the wilds of Baku and with my enthusiasms for the pleasures of Petersburg with its rich prerevolutionary past. It is to Nancy that I dedicate this book.

Robert W. Tolf

Boca Raton, Florida
March 1976

Illustrations

Immanuel Nobel at age 69.

Ludwig Nobel in the
last year of his life.

The Nobel home and Ludwig Nobel factory in St. Petersburg, 1912.

H. P. Crusell

K. W. Hagelin

Emanuel Nobel

H. A. Olsen

M. M. Beliamin

Directors of Nobel Brothers—1904

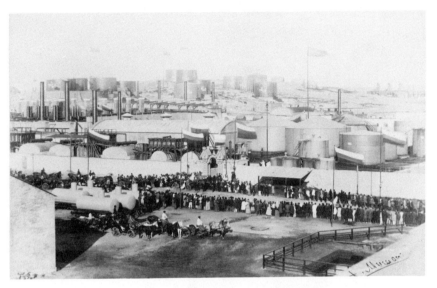

Visit of Tsar Alexander III
to the Nobel installation in Baku, 1888.

Emanuel Nobel *(center)* at age 50
with some of his foremen and workers in Baku.

Karl Vasilyevich Hagelin
at Villa Petrolea in Baku.

Emanuel's clubhouse across from the St. Petersburg factory.

Nobel Brothers' pavilion at the 1896 All-Russian
Arts, Crafts and Industrial Exhibit at Nizhni-Novgorod.

The visit of Tsar Nicholas II to the pavilion.

Harbor of Baku, 1890.

The temple of the fire-worshippers in Baku.

Ludwig and his closest collaborators in design and production of artillery weapons.

Seated: W. S. Baranowsky *(left)* and B. F. Berg. *Standing (left to right):* H. Hülfers, F. P. Kühn, J. Berson, and Ludwig Nobel.

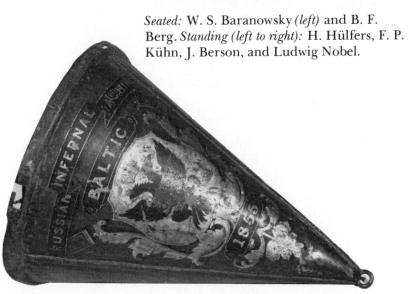

An 1855 Nobel underwater mine retrieved and decorated by the British, now displayed at the National Maritime Museum at Greenwich. *(Courtesy of the British National Maritime Museum)*

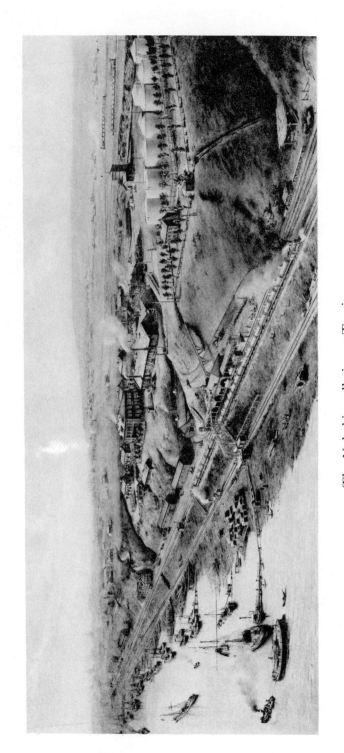

The Nobel installation at Tsaritsyn.

A Surakhany gusher.

Drilling towers at Balakhany and a reservoir pit filled with oil.

Inside a Baku oil derrick.

A Nobel oil field, 1900.

The Nobel installations at the Black Sea port of Batum.

Nobel's well
No. 25 comes in.

Destruction in the oil fields from the 1905 Revolution.

Nobel diesel generators in the Astrakhan power station.

Nobel Brothers' continuous distillation refinery.

Oil transport from the Nobel depot in St. Petersburg.

The Nobel depot in St. Petersburg.

Loading camels at the Nobel depot
in Tashkent for transport to Persia.

Nobel fuel oil moving on sleds by the Kremlin walls.

Oil transport by "arba," 1875, at Robert Nobel's refinery before the introduction of pipelines.

Bibi-Eibat, location of "the most
famous oil field in the world."

Swedish-Nobel tanker weighs anchor for demonstration run in New
York harbor with party of visitors on board. Good maneuvering and
smooth running were features of the trip.

Vandal, the world's first diesel tank barge.

Zoroaster, the world's first oil tanker.

Caspian gunboats *Kars* and *Ardagan*
powered by Nobel diesels.

Empress Elizabeth, iron barge for Volga
transport with capacity of 10,000 tons.

Nobel barges wintering at the Rybinsk depot.

Diesel tank barge *Sarmat* (*center*) and two Nobel Volga tugs, *Turkmen* and *Sart*.

The Beginnings

Nobel is associated by most of the civilized world with Sweden, but this most famous name of the north is not an ancient one nor even particularly Swedish-sounding; it dates only from the seventeenth century when a farmer, Olof, added Nobelius to his name. Adoption of surnames, usually inspired by patronymics or nearby places, was the custom of the time and Eastern Nöbbelöv in Skåne, just south of Simrishamn in the southern tip of Sweden, was the name of Olof's village. Later generations shortened the name to Nobell and then to Nobel, but did little to distinguish themselves or to provide significant indications of the genius that would later burst forth. Petrus Olavi Nobelius was a lawyer, his son an artist who painted miniatures, and a grandson an army doctor. Others remained on the farm as Nobels had done for centuries, long before they had a family name. None displayed that spark of genius. There were no brief flashes of fire, no rumbling portents of the volcanoes of creative energy and enterprise which later erupted. But there was Olof Rudbeck.

A man of the Renaissance at seventeenth-century Upsala University, Rudbeck was Sweden's first natural scientist and he made important advances in man's knowledge of the circulatory system. He founded the school of music, organized the country's first botanical garden, and composed that phantasy of Swedish romantic nationalism, *Atlantis*. Architect and administrator, scientist and musician, he combined great creative imagination with solid practical skills. He also married his daughter to Petrus Olavi Nobelius. But the Rudbeck genes of genius then lay dormant for more than a century until 1801 and the birth of Immanuel Nobel.

Immanuel had limited primary education and, as was the common custom for young Swedish males not bound to the farm, at age fourteen went to sea. When he returned three years later he apprenticed himself to a local builder who, among other projects, was erecting a triumphal arch for the reception ceremony welcoming the country's new king, Karl Johan. Immanuel enrolled in Stockholm's Academy of Art and studied architecture under the tutelage of the city's most popular master. He then attended a mechanical school where he was awarded the highest honors for his drawings and models—a wind-driven pump, a spiral staircase, storage towers, linen-finishing machinery. The year he completed his formal education he married Caroline Andrietta Ahlsell, filed patent applications for a planing machine, a ten-roller mangle, and a mechanical driving gear, and launched his career as architect-builder. He successfully negotiated contracts for construction of a bridge and renovation of several houses. But the patents were rejected and he suffered serious financial reverses in his other endeavors. When three barges of building materials destined for one of his projects sank in the deep waters of the archipelago, the Swedish career of Immanuel Nobel, Master Builder, came to an abrupt end.

Only his marriage brought him some measure of success. Andrietta was a perfect counterpoint to the demanding, temperamental Immanuel. Her humor in the face of adversity, her strength and great forbearance marked her as a true descendant of solid Småland peasant stock, a daughter from that southern Swedish province where the farmers scratched in stony soil for a bare and meager existence. Her patience, endurance—even her faith in her husband—were sorely tested during those first years of marriage as one after another of Immanuel's dreams turned into nightmares and they moved from house to house on the outskirts of Stockholm, needing but not being able to afford larger quarters for their growing family. First came Robert Hjalmar in 1829, and two years later Ludwig Immanuel. The sickly and fragile Alfred Bernhard was born in 1833 just ten months after his father had declared bankruptcy.

The previous year a fire had destroyed most of their home and possessions. Immanuel, still faced with the debts from his abandoned building projects, saw no alternative to the humiliation of declaring himself bankrupt. But he refused to be discouraged or defeated. He was determined to find practical applications for some of those ideas that never seemed to stop swirling in his head. He managed to find the backing—he could be as persuasive as he was inventive—to open a small rubber factory, Sweden's first, and he manufactured various elastic cloths as well as surgical and military supplies. He also developed an ingenious rubber backpack for the army. Fully inflatable, the Nobel knapsack was designed to be used as an air mattress, a life jacket, and as a section of a pontoon bridge.

Immanuel fabricated prototypes, submitted descriptions and cost estimates to the Swedish military, but he was unable to arouse any enthusiasm. The army was in no mood to experiment, to expand, or to spend money. He received the same negative reaction when he proposed development of a far more dramatic idea: subsurface charges of gunpowder to destroy an enemy on land or sea. Immanuel was thinking of underwater and underground mines. He had accurately assessed their defensive value and was hoping to interest the military in a research program to test his various schemes for detonation, to perfect a safe and secure means of manufacturing and then handling the mines. It was the beginning of the Nobel family's fascination with explosives.

The immediate inspiration for Immanuel's interest in the subject of mine warfare, in developing explosive charges that could be detonated from what he termed "a considerable distance" to destroy advancing armies or ships of the line, is not at all clear. Surely he was no stranger to the sea; he had spent his youth in the coastal town of Gävle a hundred miles north of Stockholm, and three years on the high seas as a young sailor. Nor was he a stranger to the army; his father had served as regimental barber-surgeon, and through his own efforts to sell various rubber supplies to the military he had been in frequent contact with both line and staff officers. As architect he had been exposed to the problems of digging into the earth for foundation construction, drilling and blasting into rock. And he may have been aware of mine warfare experiments in other countries.

Immanuel was not the first to consider the potential or to develop working prototypes. During the American Revolution a young Yale student, after an unsuccessful effort with a hand-propelled one-man submarine, assembled a number of waterproof powder-filled minibarrels, setting them afloat on the Delaware River in the hope they would explode on contact with British men-of-war.[1] It is not considered likely that Immanuel ever heard of that quixotic episode or of Francis Hopkinson's revolutionary ballad based on the event, "The Battle of the Kegs," but he might have been informed about the experiments of another American thirty years later.

Robert Fulton developed several workable mines and demonstrated them to the Dutch, the French, the British, and finally to his own country's military. Napoleon favored the idea and 10,000 francs were committed for an 1801 demonstration. It was a success, but the French military rejected further development on the grounds that mine warfare was morally indefensible. Three years later a British commission came to the same conclusion. But when Fulton returned to the U.S. he managed to win a $5,000 congressional appropriation for testing what he and the next generation always referred to as "torpedoes," an English version of the French transliteration of the Latin term for the family of electric rays. Fulton's new weapon was not utilized by the Americans in the War of 1812 against the British nor at any time during the next half-century, despite the efforts of an enterprising Connecticut Yankee, Samuel Colt, who fabricated an electrically-detonated mine controlled by an observer some distance away.[2] It was not until the Civil War that the potential of naval mine warfare was finally realized in the United States and it was then the Confederacy which showed the way, promoting at least one Union admiral into immortality—it was Farragut who gave the famous command when confronted by Confederate mines at Mobile Bay: "Damn the torpedoes, full speed ahead!"

By that time Immanuel already had perfected a series of his own sophisticated mines and the world's military strategists were awakening to the potential of mine warfare; but in Sweden of the 1830s, Immanuel was alone in recognizing the value of the new weapon and when the government failed to grant funds for research and development he was forced to put his mine ideas aside, to file them away with that scheme for inflatable knapsacks. There were other projects—Immanuel never lacked inspiration and inventiveness—and his rubber factory was proving more successful than his building endeavors had been. But financial security continued to elude him and creditors continued to hound him. In an agricultural country in which poverty was thought to be a national characteristic, Immanuel was beginning to believe that he would not have the chance to make the kind of success he felt could and would be his with the proper opportunities.

His frustration was undoubtedly not a unique phenomenon in Stockholm at the time, but Immanuel Nobel was never one to let frustration block his path or force him to a resigned acceptance of fate. When he met a visiting Russian emissary at a party in Stockholm, he was suddenly confronted by a crossroads where he saw that chance, where he found another opportunity, another new arena to seek fame and fortune.

Lars Gabriel von Haartman, a Russian-Finnish official who was later made a baron, arrived in Stockholm in the spring of 1837 to negotiate for the Russian government a trade and friendship treaty with Sweden. Haartman lived in Åbo (Turku); more, he was governor of that coastal town of thirteen thousand and head of a city commission to encourage trade and agriculture. On meeting Immanuel and listening to the young Swede's schemes and dreams he encouraged him to move to the Baltic outpost of the tsar's northern empire and to take up residence in Åbo. Haartman had important contacts in Finland and in St. Petersburg, and he promised he would introduce Immanuel to the right people both inside and outside government. He seemed as confident as Nobel that the Swedish architect-inventor would make a great success in the Finnish town.

The decision was as difficult as any Immanuel would ever make, for he had to leave Andrietta and their three sons behind. Conditions for his family in an unknown and untried area would not be any better—at least initially—than they were in Stockholm. For Immanuel, who had been struggling through a decade to make a decent living in Sweden, conditions could not be much worse. On the fourth day of December 1837, with family waving farewell on the Stockholm docks and Immanuel suppressing reluctance with hope and determination that must have bordered on the desperate, he sailed out of the harbor and headed for Finland and fortune. It was the first leg of a journey that would be taken dozens of times by Nobels in the course of the next eighty years.

The day after his arrival in Abo Immanuel called on Haartman and the governor, true to his word, saw that he met the most influential people in the city. During the next twelve months Nobel apparently justified the Finn's faith in him as he continued with his experiments and designed and built several houses, including one for the wealthy Scharlin family who had warmly received him as a lodger. The house at Nylandsgatan 8 still stands, a two-storied stone classical structure with balcony and Corinthian pillars breaking the severity of the lines. Immanuel also promoted his several rubber products and, with Haartman so well connected at court and within the hierarchy of the Russian Ministry of War, he no doubt continued to work on his mines.

This was probably the reason he decided to make another move after a year in Åbo. The Russian military was a vast state enterprise with procurement contracts running into the millions and with active programs in weapons research and development. There was no guarantee that Immanuel would ever see any of that money, but he was confident of his abilities and—with the aid of those Russian officials he had met through Haartman—he was certain of at least some measure of government interest. The decision to seek once again a new land to the east was far less difficult than the one made a year earlier.

In December 1838 he arrived in the city created by Peter the Great on the marshes of a former Swedish province. Immanuel was probably the first Nobel to set foot on Russian soil although his father's brother, Petrus Nobelius, might have been there as a soldier during the ill-fated campaign against Catherine the Great—Sweden's last attempt to drive the Russians from the sea and to reestablish its eastern Baltic outposts.

It was not too unusual for a Swede to make the short and inexpensive trip to Peter's city, to seek employment in Russia—especially in Petersburg, noted for a century as a haven welcoming pioneers and those eager to make their fortunes in trade, commerce, or industry in Russia's greatest port. Ever since Peter's decree in 1702 granting guarantees and concessions to the foreigner, the Russian window to the west had attracted adventurers, entrepreneurs, and investors. With certain changes in the ground rules, this encouragement has remained through the years a dominant factor in the economic development of the country.

The Vikings had shown the way. Their centuries-long search for plunder and commerce had swept a Scandinavian presence to the steppes of Asia, across the great rivers of Russia—the Volga, the Don, the Dnieper—to the shores of the Caspian and Black seas and all the way to Byzantium which the Vikings named the Great City and served as its imperial bodyguard. The ninth-century Nestor Chronicle relates the tradition that it was the Slavs, unable to govern themselves, who invited the men of the north—whom they called "Rus"—to rule them; those men Arab chroniclers described as "tall as palm trees, rosy-cheeked and with red hair," always armed with axe, dagger, and sword, with coarse cloaks thrown over one shoulder and with one arm always free to do battle.[3]

Near the area which Peter chose as site for his new capital the Vikings had a settlement, Staraja Ladoga, where they transshipped products for the voyage down the Russian rivers, where they rested and prepared for the labor of pulling their long ships on rollers over the watersheds and down to the Dnieper and the Volga. It was the area of major Swedish concern; the eastern Baltic was once considered a Swedish lake, the waterway to their own gates. Looking to the west and to the south for political and economic inspiration and models, Sweden could not ignore the happenings in the east, the developments in the land of the tsars. For centuries the Swedish kings—Gustavus Adolphus, Charles XII, and finally Gustav III—sought security in the east by conquering Russian armies and by erecting their own fortifications on the flanks of the empire.

From the time of the Vikings to the time of the Nobels there were thousands of Swedes who, in an endless repetition of war and commerce, fought, plundered, traded, and settled in Russia. Swedish technicians were employed by the tsars; Swedish metal workers and gunsmiths laid the foundations of an armaments industry; a Swede established the first glassworks; Swedish prisoners of war were used by Peter to cut a three-mile path through the forest for his great boulevard in the heart of his new city. At the end of that wide expanse, the Nevsky Prospekt, Peter built the cathedral and monastery to house the relics of Russia's medieval hero, Alexander Nevsky, who had beaten the Swedes five hundred years earlier. Other Swedish prisoners were sent to Siberia where they remade the capital city of Tobolsk, building a fortress and introducing the art of brickmaking.[4]

The Swedish presence and influence, the Russian remembrance of the threat, the memorials to the victories, were profound. The three hundred cannon of the fortress of Peter and Paul, that brick and marble enlargement of Peter's first humble fort of clay and wood, all faced Sweden. The city itself was founded as a bastion of defense during the Great Northern War, an enclave guarding Russia's newly-won access to the Baltic. The Golden Fountain in the grand promenade of that Russian version of Versailles known as Peterhof (now Petrodvorets) is not only a dramatic representation of the Samson-and-lion legend; it is also symbolic of Russia's victory over Sweden. Under the hooves of the Bronze Horseman, Petersburg's finest statue and the subject of Pushkin's finest poem, is a snake. The horseman is Peter the Great and the snake is Sweden.

The tradition of contact and confrontation was there, and a Swedish traveler in the mid-nineteenth century had no difficulty finding other Swedes. A Stockholm-published guidebook listed the directions to the Swedish church of St. Katarina along with information on the Russian, German, and French theaters at the Hermitage and the locations of two coffeehouses, Wolf and Beranger's by the Police Bridge and Dominique's by Peter's Church.[5] Then, too, the Swedish émigré had the comfort of knowing that he would be in a similar climate, one with short summers and dreary, relentless winters. On the city's outskirts there was the solace of the silent woods, so similar to those in his native land: groves of spruce, fir, and the beloved birch which the old Russians believed sheltered the souls of their ancestors. In the city were canals and rivers—Petersburg, like Stockholm, was built on a series of islands and was for many travelers of the time the true sister city of the Swedish capital.

For an artist, for an architect, it was an open-air museum with the monuments of Europe's architectural genius on display throughout the city. To Immanuel it was at once a postgraduate course and an overwhelming, humbling experience. How could a provincial from a small country hope to compete with the glories of design all around him? The monuments by Peter's great architect, the Italian-Swiss Trezzini, with his university, that modest Dutch-style Summer Palace, the Germanic Cathedral of St. Peter and St. Paul, the domed Cathedral of the Annunciation; the German Schädel's Menshikov Palace built at Peter's command for the former stableboy whom he made the first governor of the city; the German Mattarnovy's Anthropological Museum; the Swiss Thomon's Greek temple of a bourse; the romantic-classical style of the Russian Zakharov's Admiralty; the neoclassicism of the Italian Quarenghi's Academy of Sciences; Rossi's Senate and Synod; the baroque of Rastrelli's Stroganov Palace, the Smolny Monastery, and the Winter Palace just then rebuilding after a disastrous five-day fire the previous year; the French Monferran's Cathedral of St. Isaac and his countryman's Academy of Fine Arts; the gardens and buildings of Peterhof, designed by the Frenchman François LeBlond, chief architect of the city. It probably was not too difficult for Immanuel to determine that his future would be more promising if he forgot architecture and concentrated on minemaking, manufacturing, and development of his other inventions.

Within a few days after his arrival in Petersburg he had the chance to discuss that future, to explain his own ideas on mine warfare. At a reception given by one of Haartman's friends he overheard two of the guests discussing an experiment, a demonstration scheduled to take place the following day. One of the men was General Karl Andreyevich Schilder, the other a former professor of architecture at the University of Tartu (now Dorpat) who could have given Immanuel an exhaustive and highly educational tour of Petersburg's architectural wonders. But that particular evening the subject was the professor's work with galvanically detonated underwater mines.[6] The professor—Moritz Herman von Jacobi, a Prussian émigré, junior associate in the Imperial Academy of Sciences, and an exact contemporary of Immanuel—was appointed by the tsar in 1839 to be the scientific representative on a working committee drawn from both services to implement a program of underwater mine warfare research and development. Schilder, chief of the Division of Army Engineers and later an adjutant to the tsar, was the moving force behind the committee, and when Immanuel broke into his conversation with Jacobi to explain that he had a reliable mine that did not require observers or batteries, the general asked the confident Swede to arrange a demonstration as soon as possible.

Immanuel's demonstration took place a few days later in the Petrovsky River near Schilder's villa. There were a few anxious moments when Immanuel made the preparations, certain about his mine but doubtful about his ability to surmount the language barrier and to direct the four Russian seamen in the boat who would handle the explosive charge. But the mine was laid without mishap and exploded when struck by a small boat. The fifty-four-year-old general was ecstatic. Scarcely had the splinters landed when he rushed over to Nobel, kissed him, and started dancing. Immanuel naturally was pleased with the success of the demonstration, but he did not understand Schilder's vigorous reaction until he was told that the general lately had been under considerable pressure to find a successful underwater mine. Experiments with galvanic charges had been as unsuccessful as Schilder's submarine, that thirty-five-foot experimental model built five years earlier by Petersburg's Alexander Foundry. When Immanuel's mine was clearly shown to work the general was overjoyed. But the wheels of the Russian military colossus ground exceedingly slow and it took more demonstrations and submission of descriptions and designs—with the accomplished Immanuel doing the sketches and watercolors—before the Committee on Underwater Mine Warfare made its report. That was in September 1840, and they recommended the Nobel mine to the tsar in highly favorable terms.

Immanuel was on his way. The following year, with Tsar Nicholas standing at his side, he demonstrated his land mines, exploding a series in tandem. But the tsar and his ministers were more concerned with defending the water approaches to the empire and Immanuel was encouraged to concentrate on underwater mines, an encouragement made official in September 1841 when the tsar, through his Committee on Underwater Mine Warfare, expressed the government's approval of Immanuel's experiments. Six months later this approval was sweetened by a royal grant of 3,000 silver rubles.[7]

It was the first sizable subsidy Immanuel had ever received and it was an immediate stimulant to his mine development as well as to his other undertakings. He used the money to establish a factory, replacing a small machine shop he had been operating since mid-1839 with the backing of a group of Swedish-Finnish businessmen active in Russian trade—a venture no more successful than his previous efforts. The experience, however, was invaluable and was put to good use in his new enterprise which had the resounding title of Colonel Ogarev's and Mr. Nobel's Chartered Mechanical Wheel Factory and Pig Iron Foundry.

Immanuel had designed special lathes and other machinery for an efficient production and assembly of wheel hubs and spokes, rationalizing the process and standardizing the quality, and his inventions attracted the interest of his partner in the company, Colonel Nikolai Alexandrovich Ogarev, one of the officers involved with Schilder in mine development. Ogarev was sufficiently impressed by the dynamic and ambitious Swede, by his obvious engineering talents and inventions, that he arranged for the necessary financial support to start the joint venture. Direct involvement of the military in commercial undertakings was not at all unusual; many high-ranking officers with connections at court or in the War Ministry became intimately involved in all kinds of industries related to the military and dependent upon it. The system had numerous advantages in addition to its inherent disadvantages and dangers: official surveillance if not supervision of the manufacturing process, control of prices and quality, and —for an outsider like Immanuel—facilitating introduction to the Russian market.

Ogarev and Schilder were also interested in another new Russian enterprise: the construction of the country's first major railroad, a line running from Petersburg to Moscow. The Putilov factory was given the contract for production of some of the equipment. Schilder became a director of the company in 1848 and by then Immanuel's factory was also involved in the project. For chief engineers to plan and implement the launching of Russia into the railroad age, the government looked abroad for talent. A Russian commission studying railroad networks in the United States recruited Major George Washington Whistler, bringing him to Petersburg in 1842. The most famous American to serve the tsar since Kontradmiral Pavel Ivanovich Jones (John Paul Jones) fought the Black Sea battles of Catherine the Great, Whistler supervised construction of a locomotive factory while working closely with the Austrian engineer laying out the line precisely as the tsar had ordered—he placed his ruler on a map and drew a straight line between the two cities.[8]

Whistler organized a second factory a few miles southeast of the capital at Kolpino, designed the new roof for the city's Riding Academy, assisted in the erection of some bridges across the Neva River, and constructed new docks at the naval fortification of Kronstadt—that defensive bulwark protecting the harbor of Petersburg seventeen miles out in the Gulf of

Finland. Exciting feats of engineering, but the fame and fortune they brought Whistler were short-lived. In the spring of 1849, demoralized by the frustrations of dealing with Russian officialdom and a primitive labor force, exhausted by his exertions—and perhaps weakened by a touch of cholera from an outbreak in the city the previous year—the American major died. His body was carried by the tsar's private barge to Kronstadt where it was placed on the steamer to New York.

As the Whistler era in Petersburg came to an end, that of the first Nobel in Russia was entering its most prominent period. Once again the catalyst was provided by Immanuel's mines. Before the American had his first audience with the tsar, Immanuel—inspired by the grant of 3,000 rubles—was rushing to perfect his underwater devices. He was ready for a full-scale demonstration by June 1842. At six o'clock in the evening of the ninth a Nobel mine splintered a three-masted sailing ship skyward by the so-called Blue Bridge on the river Okhta and in the presence of the tsar's brother, Grand Duke Michael, the Committee on Underwater Mine Warfare, and a handful of high-ranking officers. Three months later Immanuel received 25,000 rubles for his invention.

That handsome payment was as much a reflection of Immanuel's achievement as it was of the degree of interest the tsar and his ministers had in underwater mine warfare. Russia was the only country in the world with an organized and systematic research and development program in the new weapon of war. Its origins went back to the time of Fulton's experiments.

Baron Pavel L'Vovich Schilling von Canstadt, a Baltic German and General Staff officer who served for a decade as military attaché at the embassy in Munich, was intensely interested in the trailblazing German research on the subject of electrical transmission of information. Familiar with the recent advances in the harnessing of electricity, the experiments of Benjamin Franklin, the work of Priestly, the early research into telegraphy, Schilling also may have been aware of the underwater galvanic detonations conducted before the turn of the century by the Italian scientists Alessandro Volta and Tiberius Cavallo. Their interest, like that of Schilling, was inspired by explorations in the techniques of telegraphy, the possibility of transmitting information over long distances by means of electric current. The Russian naturally was thinking of the tremendous strategic and tactical advantages of instantaneous military communications, but he also was cognizant of the use of an electric current sent over the same type of insulated cable to detonate explosive charges. Developing a carbon-arc fuse made from two charged bits of charcoal, he exploded powder charges across the Neva and then across the Seine while assigned to the tsar's army invading France. On one occasion Tsar Alexander himself joined the wires together to explode a distant mine.[9]

Schilling's demonstrations were more successful than those of a contemporary and fellow officer Ivan Fitstum, a lieutenant colonel in the Engineer Corps. For years he struggled to develop underwater mines fired not by galvanic ignition but by standard artillery fuses. The fact that he could not guarantee waterproofing of the fuses meant that he could not guarantee detonation, and his experiments were eventually halted by the Ministry of Marine in 1810. But record of the experiments, the reasons for the failures, were presumably kept on file somewhere in the ministry just as the successes of Baron Schilling were somewhere noted for posterity as well as for those overseeing the work of Immanuel Nobel. In postwar Russia Schilling had to turn his attention and his many talents to other fields, but he remained sufficiently interested in his earlier enthusiasms to invent the world's first electromagnetic telegraph. At his death in 1837 his place as resident authority on underwater warfare was ready to be taken by Jacobi.

During Immanuel's first year in Russia the Prussian professor tested a one-horsepower electrical engine powered by a galvanic battery to drive the paddlewheels of a small launch on the Neva. In later years he developed a system of electroplating, perfected a letter-printing telegraph and a simple single-phase transformer, while continuing to work on his mines.[10] As a member of the Committee on Underwater Mine Warfare and through his position on the Ministry of Finance's Council on Manufactures, he had regular contact with Immanuel who was continuing his work on both galvanic and chemical-contact or what were referred to as "pyrotechnic" mines. Immanuel probably experimented with some of Jacobi's mines as he continued his research and demonstrations through 1845, first in the waters near the capital, then at Kronstadt, and finally at Bjorkon outside Vyborg. Jacobi and his committee reviewed the progress, receiving Immanuel's detailed sketches, reports, and expense accounts—the Swede kept close track of every kopeck spent. It can also be assumed that the committee was consulted about the War Ministry's assignment of a young liaison officer to Immanuel's factory—the ministry was interested in more than mine production—a Swedish-speaking Finn to serve as interpreter and general aide.

This officer, later a baron and a lieutenant general, became a life long friend of the Nobel family, a confidant and business associate, a spokesman in court and—not the least important—a financial angel whenever needed. He was Carl August Standertskjöld whose presence as a full-time liaison officer was as significant a sign of Immanuel's success as were his expanded payroll, his several government contracts, his reputation as a reliable and imaginative engineer and inventor. For the first time in his life he enjoyed a certain prosperity, a comforting and encouraging feeling of success. In the summer of 1842 he sent for his family.

During his five-year absence Caroline Andrietta and her three sons had scraped out a simple existence that at times bordered on poverty. Her family was able to provide some assistance, especially in the early years when Immanuel could not send any money, but the main source of income came from a little milk and vegetable store she operated. The boys were too young to be of any real help, although during one particularly difficult time Robert and Ludwig did sell matches on Stockholm street corners. As soon as they were old enough their mother made certain they received some schooling; they were enrolled in Jacobs Preparatory, the only formal education they ever received. Robert as the eldest attended five years, Ludwig three, and Alfred only a single year before departing for Russia. Once in Petersburg they were instructed solely by tutors—Robert and Ludwig primarily in engineering, Alfred in chemistry, and all three in Swedish, Russian, German, French, and English.[11] They were also put to work in the factory, moving from one position to another, learning the business of running a business, acquiring from direct on-the-job exposure practical lessons in the problems and challenges of management, execution, and administration in nineteenth-century Russia.

Immanuel and his factory continued to prosper. He bought out his partner Ogarev and moved his plant to a more spacious building on Malaya Wulfova street, renaming the company Fonderies et Ateliers Mécaniques Nobel et Fils (Nobel and Sons Foundries and Mechanical Workshops), a title reflecting the predominance of the French language in polite Russian society and the independence of Immanuel Nobel. Tsar Nicholas, that drill sergeant of the Romanovs preoccupied with all things military, expressed official government interest in the new undertaking by sending his sons on a visit to see Immanuel's wheel-making machinery, a series of artillery shells, cannons and mortars, and of course the Nobel pyrotechnic mines.

There was no need for mass manufacture of mines at the time, although an event in Germany must have confirmed for Immanuel and the government the value of the new weapon. During the three-year war with Denmark another brilliant and busy inventor, the thirty-year-old Prussian officer Ernst Werner von Siemens, was responsible for the successful employment of underwater mines. He had been serving as chief of the Berlin artillery workshops but, like the Italians Volta and Cavallo and the Russian Schilling, he was intrigued with the potential of telegraphic communication and in 1847 he supervised the installation of a telegraph line in the suburbs of Berlin. When war broke out the following year over the issue of Schleswig-Holstein, Siemens and his brother-in-law, a professor at the University of Kiel, designed and laid a field of galvanically-controlled mines in the harbor of Kiel; the Danes were discouraged from approaching close enough to shell the port.

It safely can be assumed that the Committee on Underwater Mine Warfare levied requirements on the Russian military attaches for full and comprehensive details of the Siemens success, but whether or not the information was discussed with Immanuel cannot be determined. He was so involved with other work that he probably could do no more than discuss it anyway. Orders were pouring into his factory. He had never been so busy. His energies were devoted to the manufacture of steam engines and pipes, scaffolding, window sashes, a five-ton steam hammer, lathes, and other equipment for the workshop at Kronstadt. The government also ordered two of his large wheelmaking units and he overoptimistically established a separate factory for production of additional models. It was a typically impulsive and enthusiastic decision which proved to be as unprofitable as it was ill-conceived. In a few years he had to close it down.

More successful was the window-sash business. The magnificent Kazan Cathedral built by Alexander I on the Nevsky Prospekt to rival St. Peter's in Rome had Nobel window sashes. They were as visible an advertisement for Swedish engineering and quality control as were the dozen guns and mortars which stood for years outside the Artillery Headquarters on Liteiny Prospekt. These too were made in Immanuel's factory. His inventiveness was again demonstrated by his design of a central heating system installed initially in his own home, a neoclassical structure on Petersburgskaya Quay. A central boiler forced hot water through the house in a series of pipes. It was the first such system in Russia and he put other units into a hospital, a hotel, and several other homes.

This was a good decade for the Nobels. The three sons were showing definite signs of promise, of that genius that had coursed through Nobel veins since the days of Olof Rudbeck. A fourth son—Emil Oscar—was born in 1843 to receive a surplus of affection as the youngest Nobel; three other children had died in infancy. The family lived well and Immanuel was definitely a man of importance, a man to mark among the half-million inhabitants of the capital city. His name was on the rolls of the First Merchants Guild, and in 1853 he was presented at court and awarded the tsar's Imperial Gold Medal, a rare distinction for a foreigner.

Immanuel at last had achieved a position in life commensurate with his talents and energies. In the years following he was called on again and again to give of those talents and to devote all his energies to the challenges created by the disastrous policies of the government ruled by the reactionary Nicholas I. From the pinnacle of power as Europe's most omnipotent sovereign—even Queen Victoria referred to him as a "mighty potentate"—Nicholas was brought to ignominious defeat by his overambitious efforts to fulfill the dream of his grandmother for a southern outlet to the sea: Constantinople and the Dardanelles. But Catherine the Great's pragmatic realism would never have allowed her to be wedged into the vise which tightened on her grandson, posing as the protector of Turkey's Christian Slavs. The British, French, and even the Austrians resisted his pressures as he persisted in bringing his country to the brink of war, providing his own prediction of the outcome. "With so many tons of gunpowder close to the fire," he asked, "how can we prevent the sparks from catching?"

He certainly could not, nor could Palmerston, nor Napoleon III, nor Franz Joseph. The tsar sent his legions marching, the French and British invaded the Crimea as the Russians annihilated the Turkish fleet and then scuttled their own in the harbor of Sevastopol where General Todtleben's heroic defense was the single instance on either side of capable command. The Russian military colossus, feared throughout the continent and remembered with great respect for its victories against Napoleon, was exposed as poorly-led, ill-equipped, corrupt and inefficient, plagued by dishonest contractors shipping spoiled and inferior stores.

The navy with its paddlesteamers and sailing ships was as out of date as the army, although the Russians had steamships since the end of the Napoleonic wars—the first one was put on the Petersburg-Kronstadt run. In the 1820s a few steamships were stationed in the Baltic and the Black seas, and in 1848 the navy's first propeller-driven steamship was launched, the frigate *Archimedes*. But the process of modernization was not put on a priority basis until most of the European powers had already converted to steam and propeller in the early 1850s when the armies and navies of the continent were moving toward confrontation in the Crimea. The Nobel Foundries and Mechanical Workshops received orders for 500-horsepower engines to be mounted in three warships, the *Volya* and *Gangud* assigned to the Baltic, and the Mediterranean-based *Retvizán*, a double-decker carrying eighty-four guns and a former Swedish ship—it had been captured by Catherine the Great's navy at Sveaborg during the 1789–1790 war. The Nobel foundries forged the shafts and hammers, molded and drilled the cylinders, and supervised delivery and testing at sea. Immanuel improvised as the project developed, and his son Ludwig was intimately involved in the planning, design, and casting of the engines. The experience was important to the man who twenty-five years later would introduce an entirely new class of ship to the world. A quarter of a century after that breakthrough, the company he founded would be the first to adapt diesel engines to marine propulsion and would again make a primary contribution to modernization of the Russian navy.

While Ludwig and Robert were busy in the engineering sections of the factory, Alfred was sent to the United States to study the latest technological advances and to learn all he could about American progress in marine transport. In New York he met with countryman John Ericsson, his father's contemporary who had arrived in America the year after Immanuel reached Russia. There is no record of the subjects discussed, but the two Swedes must have reviewed current achievements in steam and hot-air technology, utilization of steam power for ship propulsion, and improvements in the screw propeller—all projects the engineer-inventor then was working on.[12] It might have been Alfred who brought back to Russia word of Ericsson's plans for ironclad vessels, if indeed he had formulated those plans ten years before the Civil War and the launching of his *Monitor*. Or perhaps it was the always-imaginative Immanuel who first conceived the idea of heavy wooden steamers reenforced and covered with layers of iron-plating. He wanted to build such a ship and to arm it with a new weapon: "flying torpedoes"—chemical contact mines propelled under water to explode on impact with a target ship. There is no evidence that the idea was ever put on paper. Immanuel scarcely had enough time to control completion of orders already in hand.

After the reliability of his first marine engines was demonstrated, he was commissioned by the Putilov Works to build additional units for the new warships they were constructing: a hundred gunboats, fourteen corvettes, and six clipper ships. With father and sons working around the clock, designing and assembling, overseeing the casting and finishing, three of the gunboat engines were completed the first year. They were installed in the *Vol, Volk,* and *Vepr.* During the Crimean War the factory delivered eleven other engines ranging in horsepower from 200 to 500. They also manufactured hundreds of mines.

The invention that had brought Immanuel his initial financial success now was responsible for a significant first in military history. Not in the major theater of operations, the Crimea—although some mines were laid in the harbor of Sevastopol—but in the colder waters of the north along the approaches to the capital, in the shallow seas surrounding the fortifications at Kronstadt and Sveaborg. Underwater mines were deployed in carefully-planned minefields as a major integral factor in defense of the harbors and naval fortifications.

But before the Ministry of War could order Nobel mines they had to hold new experiments and to have Immanuel submit new drawings and descriptions. The plans and working models developed in the 1840s and turned over to the Committee on Underwater Mine Warfare could not be found when war threatened in 1853. Detailed knowledge of the weapon had been restricted to a handful of officers who periodically engaged in jurisdictional disputes as to whether the army or navy had responsibility for Immanuel's invention. The army maintained that the underwater mine was a logical development from the land mine, that Nobel had initially been commissioned to work on land mines. In a fit of bureaucratic jealousy the army apparently refused to cooperate with the navy to foster further development. Models and plans were put aside, perhaps even destroyed by an officer of one of the feuding factions. When war threatened a decade later no one could find the papers or a prototype, and Immanuel had to assemble another on a crash basis, to test and to stage another demonstration, before being given the order to manufacture hundreds of them for defense of Russian harbors. After years of work and expense by the committee, no weapon was ready for use when needed. Not even Jacobi's plans for galvanic mines could be found. The tsar, Immanuel was later informed, never was told of this incredible bureaucratic bumbling.

The Nobel mines reviewed by the ministry and approved in 1853 were the same type developed ten years earlier: chemical-contact mines, conical-shaped two-foot containers made of strong zinc, fifteen inches wide at the top and charged with eight pounds of powder which was detonated by an ingenious slide-bar, lead-and-glass tube mechanism protruding from the mine. The glass tube, about the size of a pencil, was filled with sulphuric acid suspended over a mixture of chlorate of potassium and sugar and enclosed in an outer casing of lead which extended from the top of the mine. The mine itself was moored a few feet below the surface and held in place by sinkers and lines of rope. When the lead was bent by direct impact or by one of two slide-bars that crossed the top of the container, the inner glass tube broke and the mixture of the chemicals caused an instant heat and flame which fired the main charge.

Some of Immanuel's mines were larger, holding up to twenty-five pounds of powder; some utilized Jacobi's system of galvanic detonation and others a combination of both types with a circuitry completed only when the mine was struck by a ship. Such a sophisticated system meant day and night, fair weather and foul deterrence, a great improvement over methods dependent on individual observer-controlled firing. But Jacobi's mines were not considered to be as reliable as Immanuel's. "Nobel's mines," General Politkovsky said, "were dangerous to the enemy but also to us while Jacobi's are not dangerous to us or the enemy."[13]

Just how dangerous Immanuel's mines could be if not handled correctly was demonstrated during an early mine-laying operation when a longboat blew up killing all hands. The accident occurred just as Sir Charles Napier, hero of a half-century of Europe's naval battles and newly-appointed commander of Her Majesty's Baltic fleet, was setting sail from the English base of Portsmouth. War fever raged in Britain; there were patriotic verses and songs—"Hurrah for Old England and Charley Napier! . . . Give it to him Charley!" At Portsmouth the crowds were near delirium as the mayor proclaimed the admiral to be "one of the greatest men of his age" and urged a "great and glorious victory" to force "the Autocrat who has so wantonly disturbed the peace of the world to appreciate the courage and resources of England and France united."[14] The queen herself, standing on the deck of the royal yacht, led the procession out of the harbor ahead of Napier's flagship, the hundred thirty-one–gun *Duke of Wellington*. Victoria wrote to her uncle, the King of the Belgians, "It was a splendid and never-to-be-forgotten sight."

With such hysteria, expectations ran high, but Napier and his officers knew how ill-prepared and undermanned the ships of the fleet really were. The Admiralty hoped to remedy the situation by having Napier recruit Norwegian seamen en route—but they cautioned the commander to separate them from any Swedes they might pick up for they would never pull together in the same boat. The British also hoped to be able to convince the Swedish government to join their cause and to supply Napier with the gunboats he needed to be effective in the waters of the Gulf of Finland. Sweden reportedly had more than three hundred gunboats and Napier none, while the Russians were building them as fast as they could—some with propellers and engines from the Nobel plant. But the Swedes wanted no part in any war with Russia. They weren't even tempted by Britain's offer of the Åland islands after the successful land assault and occupation of the main fort at Bomersund. The rocks and inlets had been Swedish until 1809 but, as the king told Napier, Sweden has no desire to reclaim them.

With neither gunboats nor mortarboats Napier was unable to maneuver close enough to bombard the Kronstadt fortress which he and most military authorities of the time regarded as the strongest in the world. The shallow waters of the coast, dotted with shoals intersected by numerous skerries, prevented the *Duke of Wellington* and other ships of the line from getting into range to use their guns. Napier also knew that the waters had been mined. When he made his reconnaissance he had to send the steamers *Arrogant* and *Impérieuse* ahead of the fleet to pinpoint locations of the dangerous hidden explosives and to attempt to neutralize them. To the admiral and his men they were "infernal machines," and they had a healthy respect for their effectiveness. The British safely retrieved one off Kronstadt, loaded it with a double charge of powder, and staged their own demonstration at Riga. The mine shattered a target ship.

Napier did not think the mines were really necessary to the defense of the impenetrable fort, but he was not willing to risk his fleet to test the marksmanship of the Russian gunners or the potency of those "infernal machines" so carefully laid out by Robert Nobel—with Ludwig serving as an observer perched high atop a Kronstadt chimney. The British commander informed the First Lord of the Admiralty that "any attack on Kronstadt by ships is entirely impracticable." The judgment and the subsequent withdrawal from Russian waters were not at all well received back home where government and public were desperate for a victory. Napier's failure to "clip the Tsar's wings," as Lord Palmerston had urged, led to his immediate dismissal from command when he returned to England. "Never did an officer receive more unmerited contumely and unjustifiable censure," his biographer wrote a decade later.[15]

The Russians were delighted. The navy and the Nobels believed that their mines had played the key role in "defeating" the British in the Baltic. In January 1855 the Ministry of War signed a 116,000-ruble contract with

Immanuel for the manufacture of more mines, two hundred sixty destined for the defense of Abo and nine hundred for Sveaborg. They were to be the same type as used off Kronstadt.

In the second year of the war as another British-French fleet sailed into the Gulf of Finland, Nobel's mines proved to be a greater nuisance, tearing into the *Merlin* and the *Firefly,* causing lighter damage to the *Otter,* the *Exmouth,* and the *Vulture.* Despite the great risks involved, the British did succeed in fishing from the waters some two dozen mines. But not without mishap. While examining one of Nobel's infernal machines Admiral Seymour pushed the slide-bar across the top, demonstrating to his officers how it worked, and was nearly blown apart. He was lucky to escape with only the loss of an eye.[16]

Still vastly outgunned in Kronstadt waters, the British fleet again withdrew. It managed to dissuade the French fleet from bombarding Helsingfors but joined with it in the shelling of Sveaborg. Nobel mines did not prevent the levelling of that fortification by twenty-one mortarboats and four covering frigates. It was the last significant action of the war in the Baltic theater.

For the Nobels the war in the Baltic had been a great personal triumph, a family victory in a struggle that otherwise was marked by inept leadership, senseless and suicidal tactics, and crippling corruption. For Immanuel it was a supreme satisfaction. He knew he could improve his mines; he was certain that a more powerful explosive could be found, perhaps something similar to what two professors at Petersburg University—one had been Alfred's tutor in chemistry—had mentioned to him: the Italian Sobrero's discovery of a volatile and as yet unstable and dangerous substance, nitroglycerine. But Immanuel had no time then for experimentation with explosives. His other projects took all his energies.

1852–1857 were the golden years for Immanuel. The presentation at court, the Imperial Gold Medal, the success of his mines, the manufacture of marine engines, the steady expansion of the factory until it had a labor force of a thousand, his reputation as inventor and as one of Russia's leading engineers and industrialists were all striking indications of Immanuel's success: the ultimate rewards for his diligence, his persistence, and his genius. But problems persisted and they were mainly monetary. Despite the achievements and the recognition, Immanuel was not in a secure financial position. In describing the period a few years later, Ludwig wrote that no factory had ever before worked at such a hectic pace or displayed such great versatility. But, he was forced to add, never had such effort been so poorly rewarded. It was a lesson that none of the sons would ever forget.

Five years after the government's decision to promote domestic manufacturing in order to encourage industry to become independent of European suppliers, that policy was reversed and once again the government placed orders for heavy equipment in France, England, and Prussia. The Nobels blamed the Treaty of Paris which had ended the war in 1856; there were rumored to be secret clauses binding Russia to such purchases in the countries of the victorious allies. Others blamed the tsar. After the death of Nicholas during the war his successor, the liberal reform–minded Alexander II, wanted government revenues conserved for nonmilitary projects. But there was no oversupply of capital available for any purpose. The empire was foundering on the brink of bankruptcy. Debt mounted by the millions and the Ministry of War, largest single user of state revenues, was forced to consume its own reserve supplies as it took a 10 percent cut in its budget for 1858. The many pledges of contracts made to Immanuel by the ministers of Nicholas were ignored by the ministers of Alexander. His pleas and protests that he had expanded his payroll and factory in order to honor those contracts, actual and promised, went unheeded. By 1857 he had a surplus of labor, a surplus of supplies, but no orders. Russia, like all of Europe, was in the midst of serious economic depression.

With the government paring expenses, Immanuel looked to the civilian market for new business. He turned to the domestic steamship companies and managed to secure a contract to construct twenty marine engines for vessels working the Volga and Caspian routes. He and his sons concentrated on adapting the Ericsson screw propeller to their engines and to Russian bottoms. Important engineering breakthroughs, but the work was not enough to save the company.

There had been great economic progress during the thirty-year reign of Nicholas I. Industrial production increased fourfold, and for the first time the landowners were not the sole source of wealth and power. But as the country crawled into the industrial age the broadening of the economic base also meant that the new class of Russian—the entrepreneur, those associated with him, and those dependent upon him—were affected by periods of economic decline and depression. For Immanuel the decline was a disaster. Without government contracts he could not keep the factory running. His creditors and suppliers were demanding payment while the government refused financial compensation for cancelled orders. Immanuel dispatched Alfred, recognized as the son most adept at dealing with figures and financiers, to Paris and London to raise money to tide the company over the period of readjustment. Ludwig and Robert remained in Petersburg to work on new designs and developments they hoped would find some market.

There were other difficulties endemic to the challenge of doing business in Russia. Raw materials were extremely difficult to get. Domestic exploitation of resources was primitive and there were high tariffs on imported supplies. Free labor and investment capital, crucial to economic expansion, were not found in any abundance in an agricultural country of serfs and low tax base. The labor force was as unskilled as it was unreliable and Immanuel had to import the most important engineering and supervisory personnel, but the supply of trained and talented Swedes and Finns was not inexhaustible. Finally there was the problem of Immanuel himself, a man of massive energies and broad interests, a seminal thinker whose restless and inventive genius frequently led him to impulsive overindulgence in wildly-optimistic schemes. Less of a businessman and factory director than inventor, occasionally ill-tempered and brusque with those of lesser talent, understanding, and dedication, he was simply not the man to manage an enterprise that had grown to a size encouraged by his tremendous talents in fields other than management.

Alfred could not raise any funds; few investors wanted to risk money in Russia after the Crimean War. The government continued its policy of purchasing abroad and ignoring domestic manufacturers. The civilian sector of the economy was not large enough to absorb the heavy machinery being produced by the factory and Immanuel was extended far beyond any reasonable limit of creditors' patience. The sons saw the end coming, but Immanuel probably remained hopeful to the last. Robert, Ludwig, and Alfred would never forget the bitter lessons of their father's failure, the causes within the man and within the country. For Immanuel, then fifty-eight, there would never be another chance. In 1859 it was all over.

Twenty-two years after he had sailed from Stockholm as a bankrupt man he returned, again bankrupt. He had been to the summit and twice to the valley, and whatever hopes and dreams he carried in his breast as he and Andrietta and Emil Oscar took the long trip back to Sweden would now have to be realized by the next generation of Nobels. But he was not pessimistic about that next generation. He had no fears for the future of the Nobels. "If my sons work harmoniously, and carry on the work that I have begun, I believe that, with God's help, they will never want for their daily bread, for there is still much to be done here in Russia."[17]

CHAPTER TWO

Fame and Fortune

Immanuel's creditors turned over technical management of the factory to twenty-eight-year-old Ludwig. Married and with a son, Emanuel, he was the logical choice for the position. He had demonstrated his supervisory skills and engineering talents while working for a year on the development of new forges, for a year and a half on the mines, and for an equal amount of time in designing several of the steam engines. For the next three years he continued working on those engines, discharging the responsibilities assigned by the creditors, doing what he could to gain some measure of compensation for the cancellation of government contracts. He composed a lengthy analysis of the background, explaining the reasons for his father's expansion of the factory work force and the commitment of 400,000 rubles for new equipment—to satisfy the Ministry of the Navy's demand for higher quality and more reliable screw-propeller steam-engine units than those currently being produced by other manufacturers. Ludwig's summary was put into a formal, diplomatically-worded document addressed personally to the tsar and delivered directly to the grand duke on the second day of January 1860. But the petition was as singularly unsuccessful as was Ludwig's request for intercession by the Swedish Ministry of Foreign Affairs. In Russia, Ludwig learned, one tsar could give and another just as easily could take away. He soon was made to realize that there was nothing he could do to procure even partial compensation.

In 1862 the creditors sold Immanuel's firm to a Russian engineer who renamed it Technical Engineer Golubyev's Sampsonievsky Mechanical Workshop. But Ludwig was not without a job. With the few thousand rubles he had managed to save he opened his own factory. Initially he rented a small building on the other side of the Neva, the so-called Finnish or Vyborg side which later developed into the main industrial section of the city; then for a payment of 5,000 rubles he purchased the structure along with a little land. The neighbors of Mashinostroitelnuy Zavod Ludwig Nobel (The Machine-Building Factory Ludwig Nobel) were the Imperial Sugar Factory, and the Arthur Lessner firm founded by a Baltic German who manufactured printing equipment, machine tools, and steam engines. Lessner was a descendant of those Germans living in the provinces of Estonia, Latvia, and Lithuania who had had special privileges guaranteed by the crown when their lands were incorporated into the empire. Their prominence in the officer corps and the Petersburg bureaucracy was out of all proportion to their numbers, a fact of Russian economic and political life of great benefit to Lessner as well as to his friend and fellow pioneer Ludwig Nobel. Fifty years later their sons and heirs would join forces and a young Lessner would be placed in charge of ten thousand men on a Nobel payroll.

As the German and the Swede laid the foundations for that later success they took full advantage of the recent reversal in government policy. In contrast to Immanuel's last years in Russia, the ministries once again were beginning to encourage domestic manufactures, granting duty-free import of iron to any factory utilizing steam or waterpower. By 1866 all railroad equipment had to be of Russian make, which was a great boon to the Putilov works as well as to the Nevsk Company, a new Petersburg enterprise established in the immediate post-Crimean War years along with the Lessner Company and the Mackferson shipyards. More precisely contemporary with Ludwig's new plant were the Oboukhovsk Steel Casting and Gun Manufactory, and in Moscow, the machine shops owned by List, Perenood, and Veihelt.[1]

The greatest stimulant to this new industrial initiative was the Emancipation Act of 1861, the liberation of forty million peasants. They flocked to the towns by the thousands seeking escape into what they thought originally would be an easier life in the factories, rebelling against the act's twenty-year land redemption payments to former landlords. The population of Petersburg increased twelvefold in the forty years following emancipation, and as new and large industrial enterprises sprang from the soil, more and more freed serfs streamed to the cities and to the new jobs. Within twenty years the country had nearly two hundred fifty factories employing more than fifty-six thousand workers. It was the great era of reform, a time of basic reorganization and reconstruction, a belated awakening to the challenges of modern society and to the opportunities of the Industrial Revolution.

Nowhere was this reform more sweeping than in the Ministry of War under the leadership of Dmitry Miliutin. The Crimean War had exposed the sad state of military logistics, the overall lack of command, the shocking deficiencies in training and morale. With this specter in view and the catalyst of near bankruptcy, the brilliant Miliutin was able to force through a series of basic reforms: establishment of decentralized military districts, abolition of corporal punishment, universal conscription, reorganization of training courses transforming the huge Russian army—the largest peacetime force in Europe—into the schoolmaster of the empire.

As interested as Minister Miliutin was in administrative reforms, he was just as concerned with operational improvements, with the effect of the changes on the military effectiveness of the tactical forces and the chain of command. And he had to use many of the savings realized by his new efficiencies to improve the general morale by raising salaries and bettering the soldiers' diet while simultaneously providing the newest—and thus more expensive—weaponry. While the emperor fretted, the Petersburg press fumed, and the Minister of Finance threatened resignation, the military budget not only consumed its own economies; it gradually ground out ever-larger annual increases. A familiar and frustrating situation, one fed by its own appetite and fueled by instability abroad and technological advancement at home.[2]

For Ludwig the fresh winds blowing through the Ministry of War meant an opportunity to prove his own technical competence, to establish the reputation of his own factory. The building he took over contained small brass and iron foundries and he set about to improve their productivity and reliability. Initially he concentrated on iron. The ministry was eager to promote domestic production of chilled cast-iron artillery shells. Gruson in Germany recently had developed the process and Krupp had perfected the production. The Russians and most other governments rushed to place their orders, but they also put out bids to their local industries in the hope that Russian domestic production would make their artillery independent of foreign supply.

Ludwig was eager to make the attempt. More than a third of the state budget went to the military and he knew the potential profit from military contracts—but he was careful to avoid total dependence on such contracts, to rely on the spoken or even written word of government ministries. He had no intention of repeating his father's costly mistakes. He also knew, however, that success within the corridors of one ministry would open doors leading to other contracts. Within a year he mastered the challenge by surpassing the quality of the imported Gruson projectiles. After a period of official testing he was awarded the contract and in the next three years manufactured sixty-three thousand chilled cast-iron shells. The millionth shell came off the line in 1878. Other contracts were signed. The Ludwig Nobel Factory became the country's largest manufacturer of gun carriages, producing in a dozen years well over two thousand of them for mortars, for eight- and nine-inch cannons, and for other field pieces. It also produced the guns themselves, including one hundred eighty small rapid-fire cannons initially utilizing Gatling's design; but after 1873 Ludwig had a lighter and less complicated model of his own.[3]

Cannons and carriages were assembled on the principle of interchangeable parts, a system that was implemented only with tremendous difficulties in the Russia of the 1870s. The Russian economy was still medieval, the workers unskilled, and the education and technical training levels far below European standards. Industrialization was just dawning in the land of the tsars, and while the emancipation of the serfs did provide an instant supply of cheap and free labor, a tremendous stimulant to the development of the economy, that labor was totally without skills or training and was totally unused to the rigors and requirements of factory life and urban living.

The government had little choice but to direct if not to compel these freed serfs to take employment with one of the new enterprises then being organized on the gigantic scale that would later typify the industrial life of the empire. This sudden availability of thousands of peasants did not make the task of fine-tool manufacture or precision machining any easier; all the new labor had to be trained in fields alien to their individual lives and traditions. Who would do the training and who would do the supervising, the administrative and technical overseeing of production? The task of finding that level of talent was just as difficult for Ludwig as it had been for his father, and the solution for both men was to recruit as many Swedes and Finns as were willing to work in Petersburg. There simply were not enough qualified Russians available. Other foreign firms experienced the same difficulties and found the same solution: hiring their own nationals for positions of supervisory and technical responsibility, training the indigenous labor merely to comprehend the new work requirements which were forcing a complete revolution in their lives. The foreign colony in Petersburg, in Moscow, and in a few other cities mushroomed. Ludwig's experience was typical. As factory orders increased so did the size of the factory and the number of workers. That meant additional Swedes, Finns, and an occasional Norwegian on the payroll. He doubled, then tripled his work force as he proved to government and competition alike that he was the right man in the right place at the right time.

The continent was then engaged in that periodic and repetitive bit of insanity—an arms race. Two years after the end of the American Civil War the Prussians and Bismarck were on the march. They had swallowed a strategic bit of Denmark, defeated the Austrians in 1867, and three years later humiliated the French and Napoleon III. Every nation in Europe was rushing to prepare for new conflicts, to arm and reorganize, to convert old weapons to new, and to cast away what military planners declared to be obsolete. For the foot soldier, armed with rifle and bayonet, the new arms race meant replacement of his muzzleloader, shown in the American Civil War to be less efficient and reliable as a means of destruction and defense than the new breechloader.

The decision to progress by loading the back of the gun rather than the front led to numerous suggestions and inventions: in Britain more than fifty designs were submitted; in the United States a Congressional Board of Inquiry reviewed more than a hundred. The British, with thousands of muzzleloading Enfields in stock, selected the 1866 design of a Dutch-American wine merchant, Jacob Snider; but a few years later in that inevitable march of military obsolescence they declared the Snider breechloader inferior to the Martini-Henry weapon. In Russia, where warehouses were crammed with thousands and thousands of muzzleloaders, the government decided in 1867 to convert a hundred thousand of them to breechloaders. Nobel was awarded the contract and he utilized a design developed by a Bohemian, Sylvester Krnka, whose .65-caliber rifle—a variation of the .577 Snider—proved to be as much a stopgap as those conversions based on the Carle or Barancov systems.[4]

But for Ludwig the Krnka conversion contract was more than just a stopgap; it was an extremely useful experience which brought additional profit and fame to his factory. He was becoming a major arms manufacturer and he was keeping alive the tradition begun a good many years earlier by a Swedish prisoner of war, Ole Petersen, when Ivan the Terrible put him in charge of the Liteiny Dvor foundry to teach the Russians how to cast cannon and cannonballs.[5] Peter the Great outfitted his artillery with nine hundred Swedish cannon and Nicholas I had ordered his artillery pieces from the Swedish Immanuel. By Ludwig's time those heavier guns were being made mainly in Perm (later Molotov, now Perm again) where the Germans and Dutch had been developing an iron industry in the Urals.

Ludwig designed a new rifling machine and utilized phosphor bronze for greater strength and durability in the breech parts, machining all the bolt actions in his Petersburg plant. It was the first time bronze had been machined in Russia. When word of his successful innovations reached the Ministry of War Ludwig was given additional orders for related equipment. He designed seventy-five different lathes for fabrication of one breechloader model and for months moved from drawing board to machine shop—supervising, suggesting, inventing. Nobel machine tools were soon replacing the British product at the three government rifle armories: Tula lying south of Moscow, Izhevsk west of the Urals, and Sestroretsk a few miles west of Petersburg.

At the same time that the ministry was contracting for conversion of the old muzzleloaders it was searching the world for the best breechloader. The model finally selected was that chosen by the Spanish government in 1867, a rifle designed by a former colonel in the Union Army, Hiram S. Berdan—the mechanical engineer and marksman who had conceived the idea of organizing a group of elite sharpshooters to support regular infantry troops. With their forest-green uniforms, cowhide backpacks, and rapid-fire breechloading Sharps rifles, his marksmen—his Sharps Shooters—added a new dimension to warfare with their long-range accuracy and the rapidity of their short-range fire. The rifle he developed near the end of the war, a .58-caliber, was manufactured by Colt and the Russians purchased thirty thousand of them. But by 1870 the ministry was certain it had come up with a better weapon and, once again preferring to be independent of foreign military supply, made arrangements for domestic production.

The improved Russian Berdan known as the Berdan II or the "Berdanka" was a .42-caliber with a circular thumbpiece at the rear of the bolt and with turning-block action in place of the Berdan's lifting hinged block. It utilized a new type of cartridge, one that marked another critical phase in the development of the rifle, the first center-fire bottlenecked cartridge with an outside Berdan primer. The imaginative colonel had developed the primer for use with a brass case and at the same time devised a new method of quickly and inexpensively drawing the brass, thus helping to usher in the age of high-powered small-bore rifles with greater accuracy, range, and with a low trajectory.[6]

Responsibility for manufacture of the Berdanka was assigned to one of the government armories, that in Izhevsk under the command of Peter Alexandrovich Bilderling. Each of the three armories supplying the more than six hundred thousand troops had a general in charge and a staff of four regimental officers along with one hundred fifty permanent armorers.[7] Earlier in his career, as a captain in the artillery, Bilderling had been a fairly close friend of Ludwig's, and when he received the order for production of two hundred thousand Berdan II rifles he proposed to the ministry that Ludwig be brought into the project on a fifty-fifty basis, gain or loss, with the Nobel factory supplying the machine tools, the engineering expertise, and the administrative and financial assistance. The government readily agreed. Ludwig had clearly displayed his credentials; he was the logical choice. But the ways of the bureaucracy more often than not waxed strange and it was undoubtedly to Ludwig's benefit that one of his "best, oldest and truest friends" was in a key position to influence the decision.

Carl August Standertskjöld, the Finnish liaison officer once assigned to Immanuel, was that friend and he was chief inspector for the rifle and ammunition factories. He previously had served as head of the armories at Izhevsk and Tula and was on his way to promotion to lieutenant general. He was also a partner with Ludwig in an iron mine north of Lake Ladoga in Lupikko, Finland. The English industrialist Alfred Hill had developed a small mine in that area and Ludwig and the general—each wanting a ready source of iron ore for their factories, Ludwig in Petersburg and Standertskjöld in Tula—had invested more than 175,000 rubles in the project, two-thirds of the sum coming from Ludwig. But despite the available quantity of ore, its extraction was too difficult; after a peak production year in 1870 the annual volume declined and the company was dissolved seven years later. The total yield of pig iron was just over two thousand tons.[8]

Ludwig could be confident of the wholehearted support of both Bilderling and Standertskjöld. In the early 1870s the latter had loaned Ludwig his securities to use as collateral for bank credits to expand the factory, to construct additional buildings, and to purchase and install new machinery. He had to pay 5 percent interest for the securities and 6 percent on the borrowed cash, but these were bargain rates in a country where bank loans were not easily procured for capital improvements. The Russian general was often called upon for such loans, and until his death in 1885 he was always willing to oblige his friend. Such support and encouragement was a most valuable asset for Ludwig's plans for the Petersburg factory, the Izhevsk armory, and for other more vast industrial enterprises. His influence also made it easier for Ludwig to avoid many of the pitfalls that plagued the businessman in Russia, those problems the English director of the huge Chepelevsky Iron Works condemned as "the corruption inseparable from Russian official life." The Englishman complained that "Even when a manager is capable and anxious to do his duty, he is baffled by the process of red-tapeism."[9]

The site selected for the new government armory at Izhevsk was some twelve hundred miles from Petersburg and about halfway between Kazan and Perm; it was eight hundred miles from the nearest railroad and without telegraph. An isolated, primitive land with an abundance of freed serfs, it was an ideal challenge for someone with Ludwig Nobel's vision, for an entrepreneur, for what the times were beginning to recognize as a "captain of industry." In the course of the eight years from 1872 to 1880 Ludwig created in this wilderness a modern production facility filled with machine tools and assembly lines. He enlarged and modernized the local steel mill, increased productivity, introduced quality control, and made Russian rifle production independent of Austria which previously had supplied the raw materials. Rifles coming down the line numbered 453,455 and an additional 300,000 barrels were delivered to the Tula and Sestroretsk armories. Production cost per rifle was reduced from 27 to 21 rubles in a reversal of the suppliers' usual formula, and the lot of the workers was improved tremendously.[10]

Precedents were already well established in Petersburg where Ludwig had abolished the standard system of fines along with the practice of paying wages in kind rather than in cash. In addition to eliminating those iniquitous methods of worker suppression he put all supervisory personnel on a salary system to remove the temptation to pocket bonus payments intended for the workers, and he made certain that everyone in the plant was paid promptly and on a regular schedule—not too common a practice in nineteenth-century Russia where workers were forced to endure the worst of the abuses characteristic of the early stages of the Industrial Revolution. Nowhere in the world were those abuses worse than in Petersburg which had the oldest and most concentrated groups of workers. The filth, the "special Petersburg stench," the pitiful shacks, the degradation of life in the cupboards called rooms were vividly recorded by a son and chronicler of that city, Fedor Dostoevsky, but not even his descriptions did justice to the appalling conditions in some of the factories, especially those with workers' barracks. Cattle would have resisted the cramped quarters.

Ludwig always made certain that his workers' accommodations were adequate and well maintained. This was especially important in Izhevsk where new housing had to be constructed for the hundreds of freed serfs who streamed into the area. He encouraged them to save a portion of their wages and established a savings bank, regularly adding substantial sums from his own profits. He refused to employ child labor and was an important and enthusiastic supporter of the movement to eliminate the hiring of workers who were less than twelve years old, giving his full endorsement to the 1881 government measure banning such labor. He reduced the workday from the usual twelve or fourteen hours to one of ten and a half hours, and instituted the first profit-sharing plan in Russia if not in the world. He also started a series of free educational courses for the workers, and at Izhevsk he organized an educational fund into which the employees put one percent of their wages to help finance a technical college; Ludwig paid for the physical plant.

Decades before the business world discovered that overall efficiency and productivity are promoted by a generous and concerned attitude toward employees Ludwig Nobel was doing for the Russian industrial establishment what Alfred Krupp was doing for the German: creating a model for others to emulate. He enjoyed standing by the Izhevsk church to watch the workers coming to the services wearing shoes. They had been barefooted and in rags before they came on his payroll; now some of their women could even afford fancy dresses and parasols. Paternal, of course; but in Russia, Europe, or America in the 1870s how many new captains of industry were even paternal?

Ludwig of course made a profit on the Izhevsk enterprise just as he did when supplying other material to the military, but it was honest profit and there was none of the peculation that seemed characteristic of those in the business of supplying the ministries. Dmitry Miliutin and the government recognized this fact; in 1875 Ludwig was awarded the Order of St. Anne Second Class, the Order of the Imperial Eagle, and an official proclamation of appreciation from the Ministry of War.

But Nobel's rifles and Miliutin's reforms were not enough to prevent the embarrassment and defeat of the Russian army during the 1877–1878 renewal of that ever-simmering Russo-Turkish rivalry disputing land and seaports on the Black Sea, the protection of religious and national minorities, the balance of influence among the Balkan nations. As always, the British and Austrians nervously watched developments and this time they were joined by Bismarck. The eighth eruption of the war between the two empires lasted only nine months and the Russians—by the sheer weight of numbers fighting an enemy hampered by a system at least as inefficient and corrupt as their own—finally emerged the victors, but not until the valiant Turks had inflicted crushing defeats on the tsar's forces at Plevna in Bulgaria.

With a superiority of two to one in manpower as well as in artillery the Russians made a rapid advance to the Danube. Then Osman Pasha invested Plevna with thirty thousand of his ragtail troops, those irregular Ottoman horsemen known as the "Bashi-Bazouks." The Russian High Command with the tsar's brother in charge made another of those incredibly costly mistakes in tactical judgment that typified the war: the decision not merely to contain Plevna but to capture it. Three times the massed Russian infantry attacked and three times it was repelled with heavy losses. When twelve thousand troops were sacrificed in the last wasted effort General Todtleben, the hero of Sevastopol, was given command. Recognizing the futility of continued offensive action he ordered a classic siege; in three months the Turks were starved out and forced to surrender. Bilderling, a major general in charge of the siege artillery, was wounded and awarded the tsar's Gold Saber for his bravery.

Ludwig's son Emanuel was present at the capitulation. He had just finished a period of apprenticeship and was sent to Plevna with special steel shields constructed by the Petersburg factory and intended to protect the Russian infantry from the withering fire of the Turks. The shields served their purpose well and the young Nobel, always proud of the fact that he had been at Plevna, was awarded the St. George ribbon even though his role in the war was strictly that of supplier and observer—as his father during the Crimean War had perched on a Kronstadt rooftop to watch the British fleet maneuver around Nobel mines.

Those mines were far more effective than the Nobel Berdanka rifles. The many military observers on the scene excitedly reported to their governments that Plevna was proof of another basic change in warfare. The valor of the Turkish defenders and the suicidal courage of the Russian infantryman—and the monumental stupidity of the Russian High Command—demonstrated for the first time the unquestioned superiority of the repeater rifle. The Russians with their converted Krnka and their new Berdanka, both single-shot, were no match for the Turks armed with the Winchester 66 repeater and with the superior single-shot Peabody-Martini rifle, also American and accurate at a range of seven hundred yards—against the massed Russian infantry probably up to a thousand yards. The Turks picked off the attackers before the latter were in range to fire their own weapons. As the Russians came closer the Turks took out their Winchesters and decimated the Russian ranks with a rapidity of two shots a second from eighteen-cartridge magazines.[11]

While the Russian Ministry of War had been making its decision to convert muzzleloaders and to manufacture single-shot Berdankas, the Turks were purchasing forty-five thousand Winchester repeaters and five thousand of the shorter Winchester carbines. That was shortly after the six thousand Winchesters sold to France had provided the outclassed French forces with their only victory against the Germans. Eight years after that Franco-Prussian War the French rearmed part of their army with Kropatschek repeaters, but it was the Austro-Hungarian army which became the first in the world to adopt officially the repeater rifle as its standard infantry weapon.[12] The Americans were fond of calling the Winchester the gun you wound up on Sunday and fired all week long and later generations would refer to it as "the gun that won the West." But for the Turks, the Russians, and European military tacticians it was "the gun that won at Plevna," the weapon that ended forever the era of the single-shot rifle.[13]

The Treaty of San Stefano ended the war but not the peace. A few months later the European powers, led by the English and Prussians, gathered at Berlin and managed to diminish Russian gains and damage its prestige by trimming Bulgaria and the Russian influence in that area while placating the Turks. But the tsar still came out of the struggle with an army intact and hopefully a little wiser, with some territorial gain, and with an increased commitment to continued involvement in the insoluble problems of the Balkans.

The war with the Turks was over but not the threat of war with other nations, and for the Russians there was continued need for armaments. That, of course, meant additional orders for the Nobel factory. In the next few years, continuing to expand, the firm produced three hundred fifty thousand four-pound shrapnel shells, more than eighty thousand nine-pounders, over two hundred thousand grenades, and another four hundred fifty gun carriages, along with dozens of machines for the manufacture of powder and cartridges, and several steam boilers and cisterns for the distillation of drinking water crucial to General Skobelev's troops as they maneuvered in the Transcaspian.[14] Ludwig continued to work on the underwater mines, improving on the idea that had first brought his father to the east. He also continued to keep before him those reasons which made his father leave the east; he closely monitored government developments as well as those in his own factory, making absolutely certain he did not overcommit himself in either labor force or plant capacity until military orders were signed, sealed, and delivered to his office. In addition he promoted products in the civilian sector, procuring orders for fabrication of steam hammers, hydraulic presses, wiredrawing frames, and machine tools.

The single most important nonmilitary products of the factory were the axles and wheelrings for carriages. Ludwig had learned the technique in his father's factory but it was not until 1865 that he improved on the method of production and commenced manufacture. In the mid-seventies he enlarged and modernized the wheel-and-axle section of the plant, again personally improving the method of assembly. He designed the machinery used to form the wheelrings and invented a process for hermetically sealing the hubs of the axles to protect them from sand and dirt and to reduce the need for lubrication. He also developed special machinery for mass production of these sealed units, assembled from interchangeable parts and easy to repair or replace. The units were as reliable as they were durable. A wheel and axle which could withstand Russian roads deserved to be famous. Even in Petersburg most of the roads were cobblestones on soft foundations and pitted with too many holes, although there were a few main thoroughfares paved with wooden or stone blocks or with asphalt. But outside the city the roads were so bad the coach drivers sat sideways with their legs hanging out so they could slip off easily when jarred by a hole larger than usual.[15] Bogs, quagmires, spring thaws, rocks, potholes —these were the testing grounds of the Nobel wheel.

The result was a monopoly of the market. The Nobel wheel was the Michelin of the time, famed throughout Russia as well as abroad. Within the empire sales were organized from depots in Petersburg, Moscow, Odessa, Simferopol, Baku, Tiflis, Siberia, Poland, Finland, and several European countries. Recruitment of the sales force, establishment of sales districts, the administrative and financial direction—all were planned and implemented in Ludwig's office on the second floor of his home just in front of the factory on Sampsonievsky Prospekt. He was as interested in these administrative details as he was in the engineering, design, and processes of manufacturing. A fanatic for work, he was constantly in the plant to check on the operation of some machine he had designed, to supervise installation of a new lathe, to oversee repair of a press. There was no facet of production or distribution or direction that escaped his attention. He was the personification of Chekov's praise of work, of that declaration by one of the "Three Sisters" that "all the purpose and meaning of his life, his happiness, his ecstasies" lie in the toil and sweat of the brow of the man who works.[16]

The success of the Nobel wheel earned Ludwig his second Order of the Imperial Eagle in 1882 at the Great Russian Exhibition where he was presented with the honor for his "excellent fabrication of axles and wheelrings of Russian iron, making further import from abroad unnecessary." Once again he was applauded for his Russian production, for helping to make the Russian economy independent of foreign imports, be they finished or raw. He also had demonstrated that whatever was produced in the Nobel factory was worthy of comparison with any competing product from abroad. The honor and recognition came on the twentieth anniversary of his factory, but by that time Ludwig Nobel was concentrating on an entirely new enterprise in the far-off Caucasus on the shores of the Caspian Sea. He was focusing his energies and genius on the ancient city of Baku.

CHAPTER THREE

A New Nobel Enterprise

Robert, eldest of the brothers, was the first Nobel to cross the Caucasus and to visit Baku. He arrived there in March 1873 after visiting Alfred in Paris. By that time Alfred had a dozen dynamite factories strategically scattered around the globe and was in the process of purchasing a magnificent Paris town house on the Avenue Malakoff. His success, his newfound affluence, like that of Ludwig, was a striking and painfully obvious contrast to Robert's own career.

In 1859 when Immanuel left St. Petersburg as a broken and bankrupt man, Robert and Alfred had taken lodgings with a General Muller and for a time were involved in their brother's management of the factory. But Ludwig alone was capable of handling those tasks and by 1860 Robert was busy with a variety of other activities designed to provide a secure income but destined to force greater frustration. He worked for a time on the restoration of the Kazan Cathedral as his father had done, and then on the reconstruction of the steamer *Kryloff*, a small passenger ship cruising in Petersburg waters. Later he turned to production of fireproof bricks; he had discovered a deposit of clay outside of town and attempted to exploit the material for pressing into terra cotta. But that too was a short-lived and unprofitable endeavor and by 1862, when Ludwig was planning to open his own factory, Robert was ready to move to Helsingfors with his bride, Pauline Lenngren, daughter of a fairly wealthy Finnish merchant. With his father-in-law's backing he started another brick factory, this one in Finland, and the following year became a partner in a Helsingfors firm engaged in the import and sale of Russian-produced lamps and Russian kerosene. Competition was severe, the quality of his products poor, and the prices too high. Under Robert's leadership the company finished its first year in the red.

Alfred, in the meantime, had listened to his former chemistry tutor in Russia—Professor Nikolai Zinin—who had told Immanuel in 1855 about

Italian experiments with nitroglycerine and had recommended it for use in making more powerful land and sea mines. But Immanuel, although interested, had no time for such research and after he returned to Sweden his son Alfred tried his hand at harnessing the volatile material. In the spring of 1862, with Robert and Ludwig observing, he successfully detonated an underwater explosion in a canal near the Petersburg factory. But it was in Sweden where the first real breakthrough occurred. Alfred had moved there in 1863 and had taken out patents on the new explosive. He was not the first to experiment with nitroglycerine but he was the first to combine successful experimentation with a genius for business organization and financial management. He himself dated the era of dynamite as opening in 1864 when he set off a charge of pure nitroglycerine with a minute amount of gunpowder.

While Alfred was searching for ways to render the chemical harmless during handling, Immanuel continued to work with mines and to follow paths which had been familiar two decades earlier: trying to interest the Swedish military in his inventions. He applied for Swedish patents on land and sea mines and then on a new type of cannon. He wrote newspaper articles on mine warfare, lectured to the Officers' Society, organized demonstrations, and published elaborate bound studies richly illustrated with his own detailed sketches and watercolors: *Inexpensive Defense of the Archipelago without Great Sacrifice of Life, New Land Defense System for the Fatherland,* and *A System of Maritime Defense for Channels and Ports without Expensive Fortifications and with Conservation of Manpower.*[1] In writings and in speeches Immanuel told of the Baltic War, the blinding of Admiral Seymour, the effectiveness of the Nobel mines in preventing the British and French from bombarding Kronstadt.

He admitted that his mines were not very dangerous, that they were not as great a threat as they appeared or were feared to be, that they should have had three times the amount of gunpowder. As proposed to the Swedish military the deficiency was remedied in his new mines which were larger, more powerful, and of both contact and galvanic types, the latter no doubt inspired by his long association with Jacobi in Petersburg. For land mines he provided detailed sketches and descriptions of the needed tools for drilling holes to contain the mines. And for both land and sea mines, contact and galvanic, he considered using nitroglycerine as a more powerful explosive. His son Alfred had shown its effectiveness in the 1864 demonstration.

That was the year when Alfred's brother Emil Oscar and four companions were killed in an explosion that totally destroyed the laboratory, terrified local authorities who promptly banned any future Nobel experiments near their new home, shattered the nerves of the father, and brought the Robert Nobels to Stockholm. Immanuel suffered a paralyzing stroke a month later, but neither he nor Alfred took Robert's advice to "quit as soon as possible the damned career of inventor, which merely brings disaster in its wake."[2]

With the support and inspiration of Andrietta and his sons Immanuel survived another eight years, recovering his strength sufficiently to compose and publish a remarkable booklet which offered a practical solution to the poverty, unemployment, and emigration that plagued his native land. His imagination and inventive genius not yet stilled, he was sixty-nine when he proposed the establishment of an industry to utilize wood scraps and shavings to produce what we now know as plywood; it would be used, he explained, in the manufacture of prefabricated houses to be shipped to underdeveloped countries, to areas such as Egypt where barracks-type housing was needed for the thousands then digging the Suez Canal. Immanuel included designs for the needed machinery.

Disregarding his own advice, Robert made an agreement with Alfred during his 1864 visit to patent and then to manufacture nitroglycerine in Finland; the rights remained in Alfred's name but the profits would be Robert's. Returning to Helsingfors with renewed hope of fortune, he opened a small factory in Fredriksberg just outside the city. But nitroglycerine like kerosene failed to stimulate his interest or to provide adequate income and in May 1866 he was back in Sweden as salesman and business manager for the Nitroglycerine Company of Stockholm. The annual salary of 6,000 crowns was handsome and he worked hard at his job, but his dissatisfaction continued and he returned to Petersburg in 1870 to be put on Ludwig's factory payroll.

It was bitter irony that the most penurious of the brothers had such difficulty in finding his financial place in the world. The standards set by two such overachieving brothers would be hard for anyone to match, but it was especially painful for Robert who seemed to suffer from frequent mental malaise resembling a persecution complex, often complaining and seeking revenge for one or another insult by family, friend, or business associate. Ludwig and Alfred were acutely aware of these failings and did what they could to help him find himself as well as financial security.

But when Ludwig asked his brother to join him in 1870 it was not merely from family loyalty or charity; he needed his supervision at the factory while he took an extended holiday from the rigors of Russian weather and the demands of his many responsibilities. He had just completed the first phase of the rifle-conversion project and had finished the plans for cartridge-producing tools for the government arsenals. He was physically exhausted; his chronic bronchitis and occasional asthmatic attacks had weakened him to the point where he had to escape the cruelties of the Petersburg winter. And he also would be on a wedding trip. His first wife Mina had died the previous year and in October Ludwig married twenty-two-year-old Edla Constantia Collin. Leaving Robert in charge, he and his wife within days were en route to the Riviera which Ludwig knew well as did most wealthy Russians, just as he knew those other popular spas in Ems, Soden, Weissenberg, Modum, and the towns and villages of Switzerland—Lucerne in particular, which was on the nine-month itinerary. He returned in 1871 just in time to commence negotiations with the government on the eight-year contract for production of the Berdankas.

Robert had already tired of Petersburg and its people, complaining that he could never learn to live in such a swamp. But he remained long enough to speculate on the purchase of iron for use in constructing a tunnel near Sevastopol and on raw glycerine from Kazan to be processed into dynamite for use in blasting the canals then being dug in the Petersburg area. Both schemes failed, costing him so much money that he was quite willing to accept another offer from Ludwig who was just beginning the Izhevsk armory production and wanted Robert to study European manufacture of riflestocks and to learn from European armories the type and source of their wood, its reliability and method of handling in the factory. After that bit of industrial espionage he was to proceed to Transcaucasia to conduct a survey of the local walnut trees and to determine if they could be commercially exploited for riflestocks.

Robert executed his assignment competently, gathering data on Austrian, Swiss, and French rifle production, learning from Alfred in Paris all that he knew from his frequent contacts with military ministries around the world. From Paris Robert went to Lucerne before returning to Russia. In March 1873 he was heading for the ancient trading port of Baku, crossroads of East and West, a commercial center of great importance for Russian textiles, sugar, and manufactured goods, and for Persian fruit, silk, and carpets. If local walnut was to be found in a marketplace it would be found in Baku. There he could learn about its sources, its availability, and could make arrangements for its processing and transport. Ludwig had given him sufficient funds for initial purchase and shipment.

The Baku that greeted Robert had changed little in the fifteen years since Alexandre Dumas found that "entering Baku is like penetrating one of the strongest fortresses of the Middle Ages." It was a city unique in Russia. The Black Sea was at least partly European but the Caspian and its chief trading port were Asiatic, essentially Persian. Dumas marveled at "the fish in its rivers, the tigers, panthers and jackals that roam through the whole province; the plants; the insects—locusts, scorpions and poisonous spiders, lice and fleas; the soil and the very air of the place; all are so different from what we know in Europe."[3]

Isolated from the currents of Western thought and history by Europe's highest mountains and joined by the waters of the Caspian with the Turks to the west, the desert steppes to the east, and the Iranian tableland to the south, the region spawned conquerors and conquered, fought over by the armies of the world since man first took up arms. Two thousand years before Christ this land of the Golden Fleece had mastered the art of forging metals while Eurasia was still in the Stone Age. It was converted to Christianity five hundred years before the Russia of Kiev. The Caucasus was the crossroads of civilizations, the meeting place of East and West located between two continents and two great bodies of water leading to the river highways of Russia and Europe. Across the Black Sea the trails led to Greece and Rome; across the Caspian, to Persia and Central Asia and the trade routes to India, Turkestan, and China. In the central Caucasus the Greek gods chained Prometheus; across the foothills Alexander the Great marched his legions; and in Baku Marco Polo halted en route to Cathay and the wonders of the Orient. By then the golden age of Georgia had come to an end. Tamerlane and the Mongol hordes swept over the land and the people fled to the mountains to live for the next centuries as warring independent tribes, isolated from each other and from the world.

There was one group in Baku with counterparts only in Bombay. They were the Parsees, the fire-worshiping followers of Zoroaster. After twenty-five hundred years of persecution these believers in the age-old cult of Mithras had found sanctuary in this land of so many races and religions. Their temple of Artech-Gah with its high altar and perpetual flame—tended faithfully in the nineteenth century by the few remaining priests, the Magi—was seven miles west of Baku in Surakhany. Worshiping sun, light, and fire, their simple places of communion were put on mountaintops or near flames. In Baku their temple was in the midst of what centuries of pilgrims reverently referred to as the eternal pillars of fire (which later and more mundane generations knew to be escaping inflammable gas from fissures in porous limestone), a hundred of them burning brightly while the Parsees performed their devotions of endless chanting punctuated by clanging cymbals.

The burning pillars have fascinated man since the beginning of time and certainly ancient man knew how to utilize the oil for lighting and heating. "Thick water that burned" is mentioned in the Book of Maccabees; the Egyptians used it for embalming and valley tribes for mortar. Pliny described asphalt and a mixture of crude oil and bitumen; Thucydides recorded the wonders of Greek Fire although it was not until the seventh century that the real Greek Fire appeared on the scene—that mixture of petroleum and ground limestone which ignited when exposed to moisture. For centuries the Persians came to Baku to scrape the oil-impregnated sand near the shoreline, to load it in sacks and transport it by camel for use as fuel and light. Marco Polo reported an active export trade all the way to Baghdad. From pits of oil further inland the liquid was extracted for medicinal use. Five hundred years before Marco Polo an Arab chronicler described the Baku practice of cooking over fires of oil-soaked earth and through the centuries the oil-rich land was an increasingly important source of wealth to the ruling khans of Baku.

The fifteenth-century fall of Constantinople converted the Black Sea into a Turkish lake, and with the shores of the Caspian controlled by the Persians it was a hundred years before Ivan the Terrible reached its northern banks. After securing the Volga he captured Astrakhan. A century later Peter the Great had crossed the mountains and shown that the Caucasus could be a base for further expansion rather than a barrier, and from that time to the present the Russians have gazed covetously at the warm waters of the Mediterranean and the Indian Ocean across Anatolia and Persia, extending their empire across the Caspian and deep into the heartlands of Central Asia.

During the dozen years of the eighteenth century that Peter ruled Baku he sent several experts to study the natural resources, the potential of his newly-won territory, to confirm that old Georgian folktale that when God made the world he dropped all the minerals on their land. While Peter could not exploit the copper, manganese, lead, zinc, iron, sulphur, and coal, he did make an effort to explore the possibility of transporting Baku oil north along with the spices, rugs, and fruits regularly shipped from

Baku. But his experts decreed that the "stinking black mass" was worthless. For another hundred and fifty years Baku and its precious resource remained unknown to most of the world; the port and its polyglot population returned to their ancient ways. Turks, Persians, Tatars, Kurds, Georgians, Armenians, Jews, Russians—there was no mixture like it anywhere in the empire.

But by 1735, after Persia reclaimed Baku from the Russians, there were more than fifty oil pits being worked and the product was sold as lubricants, for illumination, and as a wondrous cure for various diseases of the skin as well as for rheumatism. When the Russians recaptured Baku seventy-five years later the oil trade was a thriving business.[4]

It was the end of the eighteenth century before the area had its first real window to the west. Clouded and imperfect as that Russian window was, it at least provided some alternative to Moslem Asia and for the subjects of Alexander I it was an introduction to the wild tribes of Daghestan. The tsar's commander and vice regent who subdued those tribes, humbling their potentate as a "shameless sultan" with "the soul of a dog and the understanding of an ass," was murdered by the Khan of Baku, an ally of the Persian shah.[5] Feigning capitulation, the khan shot the commander as he approached to accept the surrender. A futile assassination; the forces of Russia were too strong and the khan, his citadel, and the town of Baku were absorbed into the empire later that same year of 1806.

Baku was reconquered but the surrounding area remained a problem for many years. It was not until the Russo-Turkish War of 1877 that the threat from the Ottoman Empire was finally blunted and Russian hegemony over the Black Sea firmly established. But the mountain tribes remained restless and the authorities in Petersburg, for years before that war, proved incapable of dealing with the Caucasus. The misfits and incompetents of bureaucracy had been sent out to govern the new marchlands and there was the usual oversupply of mediocre jobholders who volunteered for the hardship assignment in order to gain promotion or other favors. A familiar technique and failing of all bureaucracies, this is a common last resort employed by opportunists serving all governments. Gogol understood, and he told the Russian people about it in "The Inspector General."

There were exceptions to the rule. In 1837 Nicholas visited the Caucasus and made immediate reforms, dismissing and punishing the incompetent. A decade later Prince Vorontsov arrived and for a dozen years provided enlightened leadership, constructing roads, bridges, and schools, stimulating new industrial growth. But he could not subjugate the wild marauding tribes of Daghestan just north of Baku. In the 1830s they waged a holy war against their Russian conquerors and in 1840 slaughtered nearly five thousand Russian troops attempting to put down the general rebellion. It was another twenty years before they were subdued and the areas of Daghestan, Kuba, and Karabaah brought safely into the Russian empire. More than a half million Murids then fled to Turkey, preferring the life of refugees to life under the infidel.

Pacification of the area took place when there was a growing interest in Baku and an increased realization of the value of Transcaucasia and its many resources. The production of oil was steadily increasing. By 1829 there were eighty-two pits in production; twenty years later one hundred thirty-six pits with an annual yield of almost fifty-five hundred tons. Ownership was based on a contract system with the government granting monopoly rights for exploitation in specific plots for periods of four years. The contract could be revoked at any time and there were no option rights for renewal. Contract holders therefore were never certain how long they would be able to hold their rights and they obviously were not too concerned about anything other than getting as much oil out of their pits as possible. Their contracts were expensive and the tsarist administration—never noted for its consistent economic policies and never immune to pressures, bribery, and favor-granting—was too uncertain a master. Those who leased pits understandably were not interested in measures of good management or improvement or conservation or anything else that might delay or interfere with the greatest maximum production and profit in the shortest possible time. It was a basic fact of life among the oil owners of Baku and long after the contract system was abolished, it remained the basic philosophy of most of the oil men.

By the time the system was changed it was too late; the habits were too well established, the attitudes of absolute exploitation too common. The producers knew that the contract was a total monopoly granted by royal decree and giving the holder absolute authority to do whatever he wished. Many of those granted monopolies were court favorites or military officers receiving reward for services rendered. A few were foreigners. The son of Lincoln's ambassador to Petersburg, Cassius M. Clay, held one for a time but he never utilized the rights and eventually lost them, much to the disgust of his father who later claimed they were worth millions.[6]

By 1872 the local administration managed to convince Petersburg that the contract system should be abolished. By then four hundred fifteen pits were producing over twenty-two thousand tons a year. Approximately half the oil was exported to Persia where it was used mainly to lubricate cartwheels, to grease leather and harnesses, and to provide illumination. Some of the Baku oil began to appear in other parts of the empire and there were the beginnings of a refining industry. A small refinery had been built near the Surakhany temple and in 1863 the Russian Melikov had constructed one in town. The first two drilled wells were brought in by Mirzoiev in 1871–1872 and he hit Baku's first gusher at a hundred twenty-two feet. Within a year there were seventeen wells and twenty-three refineries. The Transcaspian Company owned by Kokorev and Gubonin was soon established as the dominant force in the new industry; the company which had begun in Petersburg by promoting Russian trade with the Persians and Turcomans was shipping some of its kerosene across the Caspian and into the interior of Russia along the Volga. It was their product that Robert had tried to sell in Finland.

On 1 January 1873, two months before Robert appeared on the scene, the new government regulations abolishing the contract-monopoly system went into effect calling for public auction of thirty-three hundred fifty acres of crown lands including a thousand fifty-three acres in the administrative district of Baku and four hundred eighteen acres in outlying Balakhany where many of the pits were located. In each area where a pit already existed the surrounding land was divided into sections of 2.7 acres. Forty-eight of these sections brought a price of over $1.5 million. Chief buyers were Mirzoiev, Kokorev, and Gubonin who renamed their firm the Baku Oil Company. In addition to the initial bid price they had to pay a yearly rental of 10 rubles for each 2.7-acre section; twenty-four years later the royalty would increase to 100 rubles.[7]

By then the temple of the Parsees had fallen into a state of disrepair; there were only a few Magi remaining to tend the fires and in 1880, when the last priest died, the government issued a decree forbidding future entry of fire-worshiping pilgrims. Once the dominant religious power in Persia until persecuted and banned by Mohammed and his followers, the Zoroastrians thus were forced from still another land. The fame of Baku's burning pillars of fire now had been replaced by the fame of the pillars of oil derricks which forested the land near the Surakhany temple.

Other than the European approach to Baku across the Black Sea and then east through Tiflis (now Tbilsi) and the long and hazardous journey over the mountains, the only way for Robert to reach the city from the Russian capital was by water, taking train or passenger ship down the

Volga to Tsaritsyn (later Stalingrad; now Volgograd) and then a boat to Astrakhan where he transfered to a Caspian steamer. The all-Russian water route was faster than the round-about western approach and Robert was not the type to choose the long way around anything. His choice was fortunate. The captain of the Caspian ship was a Dutchman, De Boer, and from the very first meeting with the Swede their conversation dealt with oil. Robert was familiar with the subject from his early experience in selling kerosene and he was eager to hear from De Boer about the producing and manufacturing end of the business.

The high prices bid for sections of oil land, the long-term leases, the increasing production from the pits, the new drilling, the growing number of refineries—these were the topics Robert heard discussed when he arrived in Baku and was exposed for the first time to the swaggering optimism and greed of a boom town. Windy and dusty—the name Baku means Cradle of the Winds—primitive and polyglot, without vegetation or fresh water (the brackish Caspian could not be used for drinking or irrigation), a babble of tongues from all those now coming down from the hills to share in the new wealth. Robert was fascinated. In all his travels he had never seen anything like Baku and in all his dreams for personal enrichment and involvement he had never seen greater potential.

De Boer and his brother owned several parcels of oil-rich land in Baku along with a small refinery and they were eager to sell their holdings. They either had tired of the business or they were short of money. Whatever the reason, after Robert spent a few days in Baku inspecting their refinery, walking the real estate, and listening to the tales of other oilmen—seeing at first hand the primitive methods of extraction and processing—he made De Boer an offer. The decision was not a difficult one to make. He was impulsive but not reckless, and once he made up his mind to pursue a given course of action a strong streak of Swedish stubbornness took command and he could not be dissuaded. His offer was the "walnut money," all 25,000 rubles of it. A sudden decision taken without consultation with either Ludwig—it was his money—or Alfred, the family expert in matters of finance and investment. When he returned to Petersburg he would have to convince his brother that he had found a much better investment for the money than riflestocks.

Robert was right, but it took several years before anyone agreed with him. Ludwig, with little choice in the affair, treated the news as an obvious *fait accompli* and was resigned to the inevitability of extending further credits. Alfred took no part in the decisions or in the initial stages of setting up Robert's new business. It was all Robert's doing and his two brothers looked upon the scheme with as much enthusiasm as they had viewed his other projects. First it was fireproof bricks, then kerosene, iron, glycerine and now petroleum. What next?

After a quick trip back to Petersburg to explain to Ludwig what he had done with the "walnut money"—and incidentally to report on the lack of commercially-exploitable walnut in the Caucasus—Robert returned to Baku to take charge of his newest enterprise. He concentrated on learning the nature of the refining business, on being an oil and kerosene producer rather than a salesman of lamps and lamp oil. He was a good chemist—his brother Alfred, a rather respectable judge of the subject, said that he was a very good chemist—and in a short time he was able to suggest improved methods of refining the crude oil to produce a higher-grade kerosene than was the norm for Baku refineries.

That norm was as sophisticated as the extraction techniques: get out as much as possible in as short and inexpensive a time as possible. In refining for the all-important kerosene—no thought was given to other products in this age of illumination—the basic method was simply to burn off and distill the crude until only kerosene remained. After extraction of the kerosene, what remained was either burned or allowed to run back into the ground. Some of the more primitive distilleries consisted of a single boiler and large cauldron into which the crude was poured and heated; others were slightly more complicated and had greater capacity. But whether large or small they all belched smoke—sooty, black smoke—and the section outside town where the refineries were concentrated was christened Tchornygorod (Black Town). Baku itself became known as Bielygorod (White Town).

After studying and analyzing the refining process Robert came to the conclusion that he could do it better, but to improve and modernize the small installation he had purchased from the De Boer brothers he needed additional funds and he again turned to Ludwig, writing him that he had made important discoveries. But as Ludwig wrote Alfred, "I cannot judge these comments," adding, "I have done what I can to aid him with money, advice and technical assistance." He did not begrudge the advances; as he advised Alfred, "You and I should do what we can to help him."[8]

Robert's modernized refinery produced the highest-quality kerosene that had ever come out of Russia where the usual product was known as "Baku Sludge." And after he demonstrated the superiority of his new techniques he made plans to build a new refinery. Again Ludwig advanced the money and in 1875 the Robert Nobel Refinery commenced operation. It unquestionably was the best of the hundred forty crammed into Black Town. And it was an all-Swedish operation. In addition to Robert there were chemist Erland Theel, engineer and production chief August Avelin, and machine shop foreman Martin Westvall, who had been in Baku for several years running a little workshop. This crew was responsible for making the first breakthroughs, for achieving what in later years would be expected of anyone associated with the Nobel company: being the first, producing the most reliable product, establishing new standards, setting a pace which others had to follow, providing goals and guidelines in all phases of a new industry.

In just two years Robert Nobel proved he was the most competent refiner in Baku. His product could compete in quality with the American which then dominated the market of Russia and all other areas of the world where kerosene was sold. A decade before Robert entered the scene Russia was importing between twenty-four thousand and forty thousand tons of kerosene annually from the United States which could ship it less expensively to Petersburg or Moscow or even to Tiflis than the Russian producers could transport their own product from Baku. The government finally put a stiff import duty on the American kerosene in order to encourage their own young industry, but in those first years it was only Robert's refined lamp oil that could measure up to American oil. Three hundred barrels of his product arrived in Petersburg in October 1876. It was the beginning of the end for American oil distributors in Russia. As Nobel production soared and Ludwig entered the arena American sales steadily declined. By 1883 they were completely shut out of the market.

Robert, the pioneer in the rough and primitive frontier society, the trailblazer in the wilds of an industry that was just beginning to form in a most inhospitable land, devoted his energies to refining and did not begin drilling until April 1876. By that time Baku had nearly seventy producing wells and the area was already famous for its gushers—what the English called spouters and the Russians fountains. The world never before had seen those geyser streams of black gold shooting from the earth with a mighty hiss and roar.

The first spouting spectacular had been the Khalafi Company's great Vermishev fountain in June 1873. For four months it raged completely out of control. Millions of barrels ran back into the sands but enough was saved to plummet the price of oil from 45 kopecks a pood to a record low of

2 kopecks a pood (Russian oil was measured and sold by the pood, the equivalent of 36.114 pounds or 40 Russian pounds). Vermishev still had enough power two years later to send a nine-foot-thick column more than forty feet into the air. Vermishev was followed by "Kormilitza," the wet nurse, spouting six hundred thousand gallons a day for two years and then sixty to eighty tons a day for the next decade. In July 1874 there was another Khalafi gusher, and a year later the great Soutchastniki fountain which exploded with more than twenty-four hundred tons of oil in its first twenty-four hours. In 1876 the same company struck another at two hundred eighty feet that wasted nearly a hundred thousand tons during the three months it could not be controlled. That record was broken the following year when more than a hundred forty thousand tons from the Orbelov Brothers' gusher were lost before the well could be capped and the flow reduced to a mere four thousand tons a day.

In a small area near Lake Zabrat, soon black with oil, twenty-five gushers were brought in during a six-month period. Two of these owned by the Ararat Company yielded almost four hundred thousand tons. This was the Golden Bazaar. After the first fountain in the newer Sabunchy field there was also the Devil's Bazaar, producing at seventy to eighty feet a higher quality of oil than that found in Balakhany. As oil poured from the earth day and night the blackened workers scurried back and forth, straining to move tools, to dig trenches, shouting in a babble of tongues. It was a mad, unreal scene, something from another world. Visitors from America who were familiar with conditions in the Pennsylvania fields were stunned by the incredible waste. Watching ton after ton of oil being burned off or allowed to run into lake or sea or into the calcareous soil was a painful experience for those who only had seen wells slowly pumping a few hundred barrels a day.[9]

When word of the first gushers spread to Petersburg, to Moscow, and to cities closer to the Caspian, Baku was invaded by dozens of companies established overnight, few of them with sufficient capital to pay for more than a rig and a few laborers. They could not afford to buy a standard land parcel, a *desiatina* (2.7 acres), but they could manage to finance a *saghen* (seven-foot square). Five, ten, then fifteen rubles was the going price for ten years' use of a square *saghen* with an additional royalty of a fixed percentage of the production. As the price mounted the royalties increased and *saghens* were then split into smaller shares. The acre price in Sabunchy skyrocketed from $1,350 to $22,000 as the plots shrank to a size barely large enough to support a derrick. In the thick oil-smeared forest where vegetation long since had been destroyed and the only trees were endless rows of drilling towers, the plots became so small it was commonly said you couldn't swing a cat by the tail between the derricks. It was a raw, merciless, and often barbaric society ruled by an adventurous frontier spirit and dominated by greed and the insatiable search for riches typical of gold-rush mentality. There was a proverb among Russian businessmen that "whoever lives a year among the oil owners of Baku can never again be civilized." It was also said that anyone leaving Petersburg for Baku had his will prepared and estate set in order before departing.

With that kind of reputation, with news of all the gushers and reports of Robert's successes, Ludwig had to see the area for himself. He and his son boarded the train to Tsaritsyn, then a Volga steamer, the *Tsar*, travelling to Astrakhan and across the Caspian. The machinist on that ship was another transplanted Swede, Wilhelm Olof Hagelin, who had worked for Ludwig during the time he was managing his father's bankrupt factory and whose son would work for Ludwig in just a few years, a son who eventually would rise to a director's chair and to recognition as one of the most capable of the many Swedes who flocked to the Nobel banner through the years.

In April 1876 Ludwig and his son Emanuel steamed into the Baku harbor and for the first time looked on the forbidding barren shores, across the dry and dusty Cradle of Winds with its minarets, the eight-hundred-year-old mosque of the Persian shahs, the ancient city with its citadel and fifteenth-century palace of the khans. They were in Asia; the only evidence of a Russian presence was the gilded cupola of the Orthodox Church. For the first time they smelled oil in the air, saw the smoke rising from the brewers of that Baku Sludge, watched turbaned stevedores tie up the ship, and Armenian merchants swarm at dockside to peddle their wares. The "Nobelevski," the age of Nobel in the Russian oil industry, was about to begin.

CHAPTER FOUR

The World's First Oil Tankers

The time, the place, the setting were ideal for Ludwig's particular genius. Baku was begging for a man of sweeping, comprehensive vision able to consider the full panorama of Russian realities and economic possibilities, to temper discouragement with optimism, and to regard no barrier as insurmountable. When Ludwig Nobel entered Baku in the spring of 1876 a man with that vision arrived. It was the launching of a revolution, the real beginning of the Russian oil industry.

Although he was not a chemist he understood the value of that science in handling the raw product, the crude oil; he knew the need for constant chemical evaluation and was impressed by what Robert had achieved in that field with his new refinery. Robert's kerosene burned with the low level of odor and the high level of intensity that characterized the American product. Just as important was the favorable profit margin. But to Ludwig so much more was apparent: the tremendous potential of the area, the future of an entirely new industry, the almost unlimited markets for the products of Robert's little refinery. Realization of promise and potential could never be effected, Ludwig knew, so long as all aspects of the industry beyond refining remained bound to the traditional ways of the past and the crippling restrictions of the present. Every phase of the business would have to be examined, every area rationalized, improved, modernized, to take advantage of new techniques already in use elsewhere or those devised expressly for Russian conditions.

After his initial inspection of Baku and discussions with Robert, with his colleagues, and with some of the competitors, Ludwig went on to Tiflis where he met with Grand Duke Michael, brother of the tsar and viceroy of the Caucasus. An enlightened and competent administrator, he listened with great interest to Ludwig's enthusiastic and optimistic predictions for the future of the new industry in Baku. Again in Petersburg Ludwig researched and discussed with engineers and other technicians, preparing a series of memoranda and studies on specific ways to improve efficiency and to expand Robert's operations. He separated the studies into four main categories of pipelines, pumps, rail transport, and storage; he summarized all information currently available to him and then listed a series of questions that had to be answered before further consideration could be given to adoption of new practices in Baku. What was the exact diameter and type of metal pipe used in Pennsylvania, the kind of connectors and pumps, the size and location of iron reservoirs, the thickness of plate in what the Russians called cistern wagons, the method employed to prevent leakage of the joints? Finally Ludwig wanted to know what differences there were between the handling of crude oil and the handling of refined kerosene.[1]

When the answers began to come in Ludwig started his plans, concentrating first on moving the raw product from the wells to the refineries. The distance was some eight miles and the terrain not especially difficult, at least not for the hand-driven carts—the *arbas*—of the Tatars. These two-wheel conveyances had a large barrel suspended in the center and were capable of carrying between seven and nine hundred pounds of oil. Horse or mule provided the power, and the payment was the main source of income for thousands of Tatars. It was a half-million-dollar monopoly. And it was slow, expensive, and not always reliable; heat, storms, and Moslem religious observances interrupted the flow of oil. The costs were reflected in the price of the refined product and if that price were to be reduced the *arbas* would have to go.

Information about the Pennsylvania pipelines brought Ludwig to the conclusion that similar lines laid from field to refinery had to be utilized in Baku. But there were no pipelines in Russia and to many individuals in government and industry the very idea of sending oil through pipes over a distance of eight, five, or two miles—or just a few feet—was as mysterious as it was impractical, maybe even dangerous. The *arbas* had carried the oil for years; the *arbas* should continue to carry the oil.

Ludwig contacted the other major refiners in the hope of gaining their cooperation and cutting his own costs. He knew the project would be expensive and he wanted a united front to argue against Baku town officials who feared massive unemployment of Tatars if by some miracle the pipeline were made to work. But the other refiners were unwilling to sacrifice any part of their independence or sovereignty over their own holdings. They were jealous of their own power and position, unwilling to join in any cooperative effort, to yield to the initiative of another. Others were simply not interested in any new idea—a lack of enterprise which had harmed and held back the land for so many years. Ludwig proceeded on his own.

In Petersburg he planned the route, designed the pumping stations, supervised preparation of blueprints for the technical apparatus, and dispatched an English-speaking Russian engineer, Alexander Bary, to Glasgow to order the necessary supply of pipe—none was available in Russia. Consulting closely with his engineering department, Ludwig worked out delivery and installation schedules and exerted what pressure he could on the government in Petersburg to overcome official opposition. The Baku authorities had refused adamantly to permit pipe to be laid across their land; with most of the area between field and refinery a no-man's-land belonging to the government, for a time that proved to be an effective barrier. But Ludwig's pressures and determination triumphed over these opponents of change, and when the Ministry of Commerce signaled its approval Ludwig started implementation.[2]

The first shipment of pipe arrived in 1877. Work began at once on digging the six-foot trench which started at a central point in the Balakhany field, some two hundred feet higher than the refinery terminal in Black Town. A Nobel-designed-and-built 27-horsepower steam pump added to the gravitational pressure, pushing the oil through the five-inch pipe at the rate of three and a half feet a second. Smaller pumps were installed on the several lines leading into the main station from the Balakhany wells. The work was frequently interrupted by protesting *arba* drivers and by the coopers who made their barrels; eight watchtowers had to be erected along the length of the line. The first pipeline in America at Pithole, Pennsylvania, had provoked a similar reaction from the teamsters and armed guards had been needed to fend off attacks and vandalism. In Baku armed Cossacks were hired to man the towers and to patrol along the ground. Their presence ended the aggressive opposition; after a few futile charges the Tatars were easily beaten off. The feared warriors of the steppes won another battle. The pipeline was safe and the area returned to what passed for peaceful routine in Baku.

Ludwig Nobel's pipeline, constructed at a cost of $50,000, was then a fact of Baku life, an innovation for the Russian oil industry and—as he had predicted—a boon to the company. The freight cost from field to refinery was reduced from 10 kopecks a pood to a half-kopeck, and for the next few years other refiners were glad to pay the Nobel Company 5 kopecks a pood—later reduced to 1.5 kopecks a pood—for the privilege of utilizing the cost-saving Nobel pipeline. It took only a year for Ludwig's investment to be paid in full, including the expenses of the Cossacks and their watchtowers.

The competition then started to build their own pipelines, frequently failing to match the Nobel standards and erecting a maze of jerry-built, leaking, and ill-fitting pipes that often were the target of the midnight marauders who switched oil from another producer's pipe into their own, diverted lines into their own storage tanks, or simply let the oil run out into the sand. Despite the peculiarly Baku reaction to innovation and improvement the system obviously was there to stay. The next year the Baku Oil Company, Mirzoiev, Lianozov, and the Caspian Company put down nearly ninety miles of pipe. But Nobel remained the leader and by the turn of the century had three hundred twenty-six separate pipelines covering close to seventy miles.

Once the pipeline was completed from field to refinery another was laid from refinery to the harbor and it was there at the water's edge that the real challenge for Ludwig commenced. He had to devise a better means than wooden barrels and sailing ships to carry Robert's high-grade kerosene away from Baku and into the towns and cities of Russia where there was a growing market for the liquid of illumination. There were also the beginnings of a market for what the Tatars called *mazut* and the Russians *astatki* the residual oil left after refining the kerosene. This crude or fuel oil certainly could not be shipped in barrels; that would price it out of the market. But if it could be transported in bulk, carried from one harbor to another in large tanks, Ludwig believed that the product would compete with the more traditional sources of fuel, especially in those areas around the Caspian and the lower Volga where trees were scarce and forests depleted. At the time Russia was importing its coal from England and a domestic and less expensive source of fuel therefore was of great interest to the government and to private industry.

When Ludwig discussed the problem in 1876 with Robert and local shipping officials he decided that bulk transport was the only solution, that somehow ships had to be designed to carry not a few hundred barrels but tons of bulk oil in the hold. There had to be a fast, reliable, economic means of moving lakes of oil from the isolated regions where it was readily available to the heavily-populated centers where it was in demand. The cistern ship, the oil tanker, was the solution. Stephen Goulishambarov, State Counsellor in the Ministry of Commerce and Industry, was in Baku when Ludwig was examining and discussing the various possibilities—as the Russian described it, when "the great idea of the transport of petroleum and its products in bulk was first conceived." It was, he declared, "the most important fact in the entire history of the petroleum industry." Goulishambarov, his country's chief oil expert and a student of the industry familiar with the world's oil economy, participated in those discussions and thirty years later wrote that "the successful solution of this difficult problem was entirely due to Ludwig Nobel."[3] It was a solution, a discovery, to rank with brother Alfred's invention of dynamite.

Ludwig took his idea to the shippers; first to the Caucasus and Mercury Company, largest on the Caspian and heavily government-subsidized, and then to other transport firms. He found little support even though the shippers were constantly complaining about the leaky barrels. Other producers were even less receptive to this concept than they had been to the idea of pipelines. They argued that if the idea was really practical the clever Americans with all their business sense would have thought of it and would have used such tankers for transport across the Atlantic. They did not know that the Americans had in fact tried one tankship but it proved unequal to the challenges of the ocean. In America wood was plentiful and inexpensive but in Russia barrels accounted for half the price of the petroleum; they had to be constructed of wood brought in from remote areas of the empire or imported all the way from the United States. Ludwig argued the facts of the case: the standard barrel weighed sixty-four pounds or some twenty percent of the oil it held, which meant that a fifth of any cargo of barrel oil consisted of wood that could be shipped only one way; returning empty barrels any distance obviously was a false economy. As more and more oil gushed from the soil of Baku the price of the raw product fell but the price of the barrels remained the same; in time the barrel would be worth more than the oil. The system had to be changed. Dependence on barrel transport was as impractical as reliance on *arbas*. Ludwig's arguments, as logical and persuasive as they seemed to him, failed to sway the opposition. There was absolutely no support for the scheme. It was too great a risk; it had never been done before.

Disappointed but not discouraged by this resistance, Ludwig proceeded on his own. His first experiments were with barges loaded with fuel oil. They leaked, but not too badly as long as the load level matched that of the water level. It was not really the answer he was seeking but he proved to his own satisfaction that the skin of the vessel could be used as the outer shell of the oil container. He then began to sketch a specially-constructed tankship, one built exclusively for bulk transport of oil. Safety considerations were the first problems to surmount: sealing off the oil from the boiler area, making allowances for expansion and contraction caused by temperature changes, ventilating the hold so no gases could accumulate. His own education in marine engineering, earned on the job during the 1850s when he worked on the navy's steam engines, provided the necessary background and enabled him to draw up general plans and to work closely with his draftsmen on the engine specifications; but for the detailed drawings, the actual blueprints, he had to rely on the firm selected to build the ship. When no Russian showed interest he went to Lindholmen-Motala in Sweden and there, with Director Sven Almqvist, he designed the world's first oil tanker.

The contract was signed in January 1878. Nine months later, when Ludwig could be certain of the practical soundness of his ideas, he ordered two more tankers of the same design. They were christened *Buddha* and *Nordenskjöld*. The first of the line, the first oil tanker in the world, was named most appropriately *Zoroaster*. It was the first of three Nobel ships honoring the ancient fire worshipers.[4] It was also one of the first ships in the world to use Bessemer steel and, like its two sister ships, utilized built-in iron tanks both fore and aft to carry its kerosene cargo of two hundred forty-two tons. One hundred eighty-four feet long and twenty-seven wide with a draft of nine feet, the *Zoroaster* had to be small enough to negotiate the long journey to Baku: from Sweden across the Baltic, through the lakes Ladoga and Onega, along the Rybinsky and Marinsky canals, down the Volga, and across the Caspian. The trip had to be made during the high waters of the spring thaw.

The Petersburg factory designed and built pumps and other special equipment and Ludwig personally was responsible for planning the series of twenty-one vertical watertight compartments which were essential to keep the ship afloat during rough weather. The Caspian with its sudden storms and squalls was as dangerous as any ocean; vessels built to withstand Caspian conditions could sail the Atlantic or Pacific.

Ludwig's design of the *Zoroaster* with the engine amidship and the cisterns in the hold joined by pipes, proved as seaworthy as it was revolutionary. From the moment in 1878 when it was put into service on the run from Baku to Astrakhan it was the marvel of the Russian merchant marine, the model studied and copied by shipping firms and brokers, draftsmen and designers, oil industrialists and government officials. Ludwig made no effort to keep secret any part of the design and he rejected the pleas of his associates that he take out patents. He did not wish to profit in that manner, to restrict dissemination of ideas that could benefit the entire industry. *Zoroaster* was open to inspection and imitation by anyone who made the trip to the Caspian or who discussed the ship with Ludwig and his engineers in Petersburg or with Almqvist and his crew in Motala. In the years after 1878 there were dozens of British, Germans, Americans, Russians, and Swedes on the scene studying the design and performance of the world's first tankers. After *Buddha* and *Nordenskjöld* came larger ships which had to be divided into separate watertight sections and transported on pontoons in order to travel the old Viking water route to the Caspian.

In November 1880 Ludwig ordered a new type of tanker which utilized the shell—the skin of the vessel itself—as the wall of the bulk cargo hold. The load of kerosene completely filled the storage area from side to side and clear up to the deck; there were two additional tanks constructed aft of the engine room, abutting on the deck and allowing just enough space for the propeller shaft. Buoyancy was provided by partitioned sections fore and aft which served as crews' quarters and pumping room. The ship was christened *Moses* and within a year of its launching Ludwig ordered seven more vessels of the same design for delivery in seven to eight months. They were named *Mohammed, Tatarin, Bramah, Spinoza, Socrates, Darwin, Koran; Talmud* and *Calmuck* followed a few months later—fitting tribute to the richness of Caucasian history and to the individual genius of man. The homeland had not been forgotten: the third tanker, *Nordenskjöld,* honored the Swedish explorer who recently had discovered the Northeast Passage, circumnavigating northern Asia from the White Sea to the Bering Straits.

The new system of transport was not without its hazards and when accidents occurred the voices of alarm joined the chorus of opponents who feared Ludwig's innovations. In 1881 the *Nordenskjöld* exploded while loading kerosene at Baku. A gust of wind caused the tanker to list and the iron pipe carrying the kerosene from dock to ship was jerked from the hold's opening. Kerosene spilled over the deck and into the engine room where a crew of mechanics was working on the engine with the aid of light from kerosene lanterns. Half the crew was killed. A few months later another Nobel ship, one of the Volga barges, burned in similar circumstances.

As cries of the critics echoed in Petersburg, Ludwig called for complete accident reports and interviewed those crewmen and officials with first-hand knowledge. He was not about to give up his idea so easily, not about to admit defeat. If there was a problem he was certain there would be a solution. He reviewed the logistics of the situation, the procedures and practices of loading and unloading, the reasons for the two fires. Soon he realized that the fault lay in the loading pipe. As long as it remained rigid there would always be a danger of spillage during bunkering operations. What was needed was a flexible pipe, a leakproof conduit with couplings that could move with the ship and remain firmly fastened to the tanker's loading hole. Design and fabrication was a fairly simple matter and Ludwig's new pipe became standard equipment on all Nobel tankers and then on all Russian tankers and on others beyond the borders of the empire. It remained a standard tanker item until replaced years later by the steel-coil reenforced rubber hoses.

The Nobel tankers averaged two hundred forty feet in length. The *Koran* and *Talmud* were the longest at just over two hundred fifty feet. After the *Zoroaster* the engines, between 100 and 120 horsepower, were installed aft. Carrying capacity averaged seven hundred fifty tons. Most Nobel tankers were built by Motala although Bergsund in Sweden constructed two of them and by the mid-eighties Russian, German, and English yards had Nobel contracts while Motala was receiving orders from Ludwig's Russian competitors. In the years 1882–1883 Almqvist and his company built five Nobel-type tankers for Hadshhi Ali Dadascheff, typical of those who had so vigorously opposed Ludwig's ideas and now rushed to imitate. In 1882 the Crichton yard in Åbo built the *Merv* modeled after Nobel's *Spinoza* but somewhat smaller and with cabin accommodations for forty passengers. Crichton also constructed the two hundred twenty-five–foot *Batoum,* and for the Kokorev company the two hundred–foot $75,000 steamer *Surakhany* which carried five hundred tons of oil. Mitchell in Newcastle-on-Tyne built a number of tankers in the early eighties, mainly for the Fedoroff and Pavlov firm. In Russia the Rybinsk-based Yuraviev Engineering Company built a two hundred fifteen–foot *Sheksna* with a capacity of five hundred twenty-five tons.

By 1885 there were eleven Nobel tankers on the Caspian and two in the Baltic, the *Petrolea* and *Pirogoff,* and Russian oil companies were spending more than $5 million on new ships to carry their precious cargoes. Ten years later nearly a hundred tankers of various sizes and shapes on the Caspian carted Baku's black gold which was then flowing from Baku refineries at the rate of one hundred twenty million gallons a year. At the turn of the century nearly a third of the Russian tankers were under eight hundred tons; the largest were those belonging to Nobel, close to two thousand tons. Loading their liquid cargo in the well-protected Bay of

Baku where several tankers could be charged simultaneously, the line of ships made a never-ending procession along the six hundred–mile route to Astrakhan. An average Nobel ship could be filled in three hours, could reach the headwaters of Astrakhan in thirty-four, unload in four, and be back in Baku thirty-two hours later. Discharge of the oil had to be done a hundred miles from Astrakhan itself because the shifting Volga headwaters were too shallow for any vessel with more than a nine-foot draft and—later, after harbor dredging—one of twelve feet. Transshipment via the fleet of barges took another thirty-five to forty hours.

During the thirty years after the launching of the *Zoroaster*, with the world entering the gasoline age, the size and number of tankers greatly increased. By 1907 there were two hundred seventy-six oil-carrying ships plying the seas to transport crude and finished petroleum from remote sources in isolated areas to the populous urban markets. In Russia nearly a hundred fifty tankers had a combined carrying capacity over one hundred forty-two thousand tons. All but eight of these were confined to the Caspian, a dramatic reflection of the Russian concentration on the internal market. Other countries worked the international routes but they were slow to follow Ludwig's lead. Early in the eighties three Norwegian ships, the steamer *Stat* and the sailors *Jan Mayn* and *Lindernoer*, each with an iron tank secured in the hold, carried American oil across the Atlantic. But despite their intimate knowledge of the *Zoroaster* the Americans and the British were reluctant to risk building transatlantic tankers. They apparently regarded the Caspian as a safer body of water than the ocean and were unaware of the sudden squalls and full-blown tempests that took the measure of any ship.

In 1883 Nobel's representative in Antwerp, Henri Rieth, convinced the British shipping firm of Lennard and Sons of the wisdom of importing Russian oil and together they worked with the Craggs shipbuilding firm of Middlesborough in converting a steamer, the *Fergusons*, to the trade by fitting tanks tightly into the hold. At the same time Nobel placed orders for the first time in other British yards. Colonel Henry F. Swan, well-acquainted with Nobel's tankers, drew up the plans for the Nobel ships *Blesk*, *Lumen*, and *Lux*, which had their holds divided by a series of transverse bulkheads with a single longitudinal bulkhead running the length of the hull; each of the compartments was fitted with its own trunkway, its own expansion tanks. These ships, like those Swan designed for other Russian companies, were transported down the Volga in sections and assembled on the Caspian.

There were other British tankers. In 1884 the *Bakuin* was launched, the first British-owned tanker built in a British yard. Alfred Suart was responsible—he commissioned most of the early British tankers. Within a half-dozen years he had a total of sixteen in service. The British were slow to warm to the idea, but with Suart and Swan showing the way British yards soon were working overtime to try to close on that commanding lead built up over a dozen years by the Swedish shipbuilders. Their interest was given added stimulus when Nobel's *Petrolea*, which had been working the Libau-Lübeck run on the Baltic, arrived in London in 1886 with a thousand tons of bulk kerosene. The Russian captain docked his ship at Regent's Canal Dock and British shipbrokers Lane and Macandrew who had commissioned the trip ran a pipeline under the streets from dockside to the Atlantic Wharf where the cargo was pumped into lead-lined storage tanks. Russian bulk kerosene on the Thames in the heart of London! It took little imagination to comprehend the significance of that precedent for the industry. In the next few years the firm of Lane and Macandrew showed repeatedly that it had that imagination as it established itself as a seminal, powerful force in the industry.

It was left to a German, Heinrich Riedemann of Bremen, to realize finally the transoceanic value of Ludwig's achievement. In 1884 his three thousand–ton sailing ship *Andromeda*, outfitted with iron tanks built into the hold by Armstrong-Whitworth of Newcastle-on-Tyne, crossed the Atlantic with a bulk cargo of oil. Colonel Swan was the designer and the following year his *Glückauf*, a three hundred–foot twenty-three hundred–ton steamer also commissioned by Riedemann and built in Newcastle, became the first real tanker to cross the ocean. But when it arrived in New York to take on its first load of oil the leaders of the barrel industry made such a threatening protest that the captain moved on to Newfoundland where he took on a cargo of coal. The next voyage was more successful, and when a load of bulk oil was discharged in Geestemunde in July 1886 a milestone was reached. Ten years after Ludwig had conceived the idea and eight years after he had given the world a practical working model of that idea the discovery was accepted and copied outside of Russia. The worldwide revolution in transportation was under way.[5]

In Germany the success of Riedemann's *Glückauf* was quickly followed by improved models as tanker keels were laid as fast as the drawings could be completed: *Vorwärts, Gut Heil, Willkommen, Energie, Minister Mayback,* and then dozens more. In five years there were nearly eighty crossing the Atlantic and in the ensuing years as the figures doubled and then doubled again and the size of the tankers increased to levels never dreamed of, with captains using bicycles to move from one end of the ship to the other and crews relaxing in topside swimming pools, the small beginnings on the Caspian Sea were forgotten. With half-million-ton tankers now plying the oceans and million-ton monsters on the drawing boards there is a whole new vocabulary: supertankers, VLCCs (very large crude carriers), and ULCCs (ultra-large crude carriers) now unload their millions of gallons at superports located miles from land. "The fruitful seed sown by the genius of Nobel in Russian soil," as a British historian wrote nearly seventy years ago, had indeed "brought forth a brilliant harvest in both hemispheres."[6]

CHAPTER FIVE

A Leader in Industry

Ludwig Nobel's tanker fleet was the most dramatic innovation in the Russian petroleum industry of the 1870s but it was only a part of the general transportation and distribution system which he conceived and organized. Pipelines to carry the raw oil from field to refinery and the finished product to dockside; tankers to carry that product six hundred miles north; smaller tank barges for transshipment; barges for moving the oil up the mighty Volga to the storage depot at Tsaritsyn with its gigantic tanks and reservoirs; railroad tankcars and barges carrying the oil from Tsaritsyn to all corners of the empire; retail distribution centers in every major area of population; a labor force of thousands from field to consumer, drilling for oil, building barges, fitting pipe, refining kerosene, repairing tankcars and ships, peddling the products—this was Ludwig Nobel's empire within the empire. He created it in ten years. From well to wick it was all Nobel.

It was the achievement of an individual who thoroughly dominated all aspects of an industry which he was literally creating as he went along, inventing this piece of equipment, supervising the layout of that particular freightyard, drawing plans for tankers and barges, for railroad cars, pumps, engines, storage tanks, warehouses; organizing sales areas, arranging bank loans, overseeing company finances. He was president, chief engineer, sales manager, an entire research and development department, chairman of the board, and market analyst. An incomparable, insatiable overachiever, his deeds in Russia of the time were without parallel. On a world scale only a de Lesseps, builder of the Suez Canal, was of equal rank.

What he achieved for the industry on sea he achieved again on land, building his own fleet of railroad tankcars. The pattern was the same: contact the other major producers in Baku and receive a negative response; contact the shippers—in this case the Griazi-Tsaritsyn Railway Company—and receive a negative response. Then take out pen and paper and do it yourself. But this time someone else had blazed the trail. He had the American iron boiler–type cars, then in use for some ten years, as a model. The cars he designed were smaller than the American ones and held ten tons or about twenty-six hundred gallons. The iron plate was initially produced in his Petersburg factory and later at the Tsaritsyn freightyards. Rolling stock was furnished primarily by Putilov.

The first hundred tankcars assembled at Tsaritsyn proved more successful than even the optimistic Ludwig had hoped and he immediately ordered additional stock. They first were put on the Griazi-Tsaritsyn line and by 1881 a Nobel train was delivering oil by rail to Petersburg. Ludwig demonstrated that land transportation, while more expensive than water, was essential. There was no other way to distribute petroleum products during the long winter when the Volga was frozen and river transportation halted, the very time when demand for illumination—and later, fuel —mounted to peaks directly proportionate to cold and darkness. The fifteen hundred tankcars that Nobel had in use by 1883 were seldom idle during those winter months; sixty trains of twenty-five cars each carrying oil to the central distribution depots Ludwig had constructed throughout the country.

For as soon as he had his first tankcars in use he asked the central railroad administration if they would participate in a secondary related project of building storage depots and terminals with the necessary sidings at key points along the railroad lines. They refused—by this time nothing but negative answers were ever expected but Ludwig never tired of asking—so he built his own. He also had to construct repair yards for his cars; the railroad had refused to service his private fleet. Thus he was forced to establish an entirely new organization: a company railroad with fifteen hundred cars, a couple of dozen locomotives, all the personnel to man and maintain the lines and equipment, the marshalling yards and sidings, repair sheds and machine shops, a storage and distribution control system. He prided himself on being able to pinpoint at any given time the location of any one of his trains and the quantities of petroleum available in each of the depots.

Those depots were gigantic. Tsaritsyn, the first, was completed in 1880 and had nine iron tanks holding five million gallons with pipelines running from wharf to tanks. By 1900 the capacity was increased to over twenty million gallons, more than a third of the Tsaritsyn total for all the oil producers. The Volga is over a mile wide at Tsaritsyn, allowing ample maneuvering and docking room, and the lower Volga at that point is usually ice-free by the end of March or early April and is then open until November ends, three weeks longer than the shipping season further north.

Similar facilities were constructed in Riga, Moscow, and Petersburg, then at Domnino (now Orel), and distribution depots at Nizhni-Novgorod (Gorky), Warsaw, Rybinsk, Yaroslavl, Samara, Saratov, Kazan, Perm, Tver (Kalinin), and numerous other sites. By 1885 there were more than forty depots, ten years later another twenty, and by 1900 a total of one hundred twenty-nine. The figure doubled in the next decade. The largest was at Domnino about seven miles from the town. Because of its central location it was selected by Ludwig in 1882 as the key storage and distribution center for all European Russia. On two hundred seventy-five acres and at a cost exceeding a million rubles he built tanks with a capacity of eighteen million gallons, six miles of track with sidings for eight hundred cars, and a large workshop. As the number of smaller depots increased and the railroad reached more remote areas of the empire the usefulness of Domnino as the central cog in the Nobel network declined and in 1900 it was closed. By then there were thousands of tankcars in use on Russian railroads. The voices of opposition that had so vigorously opposed Ludwig's ideas as impractical and dangerous—until they saw the practicality and profit and monopolistic threat and demanded that all tankcars be taken over and run by the state—were stilled. They had their own tankcars or used those of the state railways. A few years after the first Nobel trains were circling the empire the railroad administration saw the light and ordered their own tankcars.[1]

The opposition was overcome but there were annoying recurrences; just a few weeks before completion of the depot on the outskirts of town the Moscow authorities decreed that it was a menace to life and limb and had to be torn down and built a greater distance from the city. The decree was later rescinded but only after considerable effort was directed primarily against the importers of American kerosene who were not too pleased to see a new Nobel depot being erected in their territory.

Opposition never really bothered Ludwig, never really impeded his progress or diminished his determination. He was one of those individuals strengthened by it, one who seemed to thrive on it. "An industrial undertaking, properly managed and well organized," he once wrote, "involves a constant struggle. . . . Its success is dependent upon foresight, perseverance, industry and economy."[2]

Ludwig had an overabundance of all four qualities and the fruits of his "constant struggle" were bearing still more fruit: the Nobel name was spreading across the land on tankers, railroad cars, and those barges sailing the mother of Russia's waters, "Matushka" Volga, from the delta city of Astrakhan near the ancient capital of the Khazar empire, principal portage point for the Viking traders who used the Volga as their highway to the south. For Nobel the mighty Volga was a highway to the north. By the early eighties the company had a dozen iron-tank barges carrying kerosene, four fitted with iron tanks, and another twenty-eight made of wood for transporting fuel oil. Utilization of the twenty-three hundred miles of Europe's longest river was not a novel idea—the Vikings had proved that—but organization of a fleet of barges as part of a complete distribution network, as a key element in a fully-integrated industry, was new. Once again Ludwig showed the way for others to follow using barges of wood or iron. The Russians had discovered that water pressure maintained against the wooden slats of the boats kept the oil from leaking through; for lighter cargoes like kerosene deckstones were added in order to lower the boat to proper water level. That expedient apparently had eluded the Americans for they abandoned the use of barge transport for kerosene.

Motala built Nobel's first steam-powered tank barges; for years their *Kalmyk* and *Tatarin* paddle-wheelers were familiar sights along the shores of the Volga. Then the new Nobel shipyards at Tsaritsyn started construction: the *Elena* and *Elisabet* were the first and were followed by dozens of sister ships in the years to come as "Matushka" Volga carried some fifteen hundred barges moving the raw and finished products through the empire. None were as imposing as the Nizhni-Novgorod–based *Martha Poseidnetza,* one of the largest barges in the world, more than five hundred feet long with four rudders and a capacity of nine thousand tons of oil. Like the other supersize steel barges it was of light construction and had to be loaded at many places simultaneously to avoid undue strain and cracking of the metal.

In the more isolated areas of Russia, however, where the rail lines were just beginning to draw in the primitive lands and where there were no possibilities for bulk water transport, there was still a need for barrels. For these markets Ludwig and the other producers imported barrels or the wood for their making. Barrels were also used as the standard shipping container for lubricating oils. To cut the cost of their use Ludwig designed a barrel factory utilizing the most modern machinery to build these containers on the American model. Located in Tsaritsyn it could turn out a hundred thousand barrels a year. As the number of depots and terminals multiplied, other smaller barrel factories were built as part of the individual storage and distribution complex.

Basing his manufacture of barrels on the American system was similar to Ludwig's utilization of the American experience in designing tankcars or in drilling for oil. His Russian engineer Alexander Bary was making Scotland only his first stop when he was sent there to order pipe for the new Nobel pipeline. His instructions were then to proceed to the Pennsylvania oil fields and to learn what he could about all aspects of pipeline use, extraction, and drilling, and to recruit experienced drillers to work Nobel wells and to help train Nobel engineers. Ludwig was not deterred by the report of the European expert who had recently visited Baku and declared that no oil of commercial value would ever be found below two hundred feet. A noted geology professor had said much the same thing in 1863: there was no oil beyond sixty or seventy feet and the new steam-powered drills would prove to be useless. The professor favored the ancient method of hand-dug pits. By the time Ludwig's six American drillers arrived on the scene in 1878 there were no pits to be found but there were almost three hundred steam drills, an increase of three hundred percent in just two years.

The Russian method of drilling was based on the steam-powered, pole-tool, percussion system supported by fully-boarded eighty-foot wooden derricks. Solid square rods held the sinker bar, drilling stem and bit, and a set of so-called jars which picked up the bit at the bottom of the hole, releasing it again to fall in the percussion stroke. The iron drilling rods were between an inch and a half and two inches in diameter, and the well hole some one and a half yards at ground level decreasing to about half that size as drilling progressed. For the unskilled labor of the Baku fields it was a practical and popular system.

The experts Bary brought back to Russia taught the Nobel mechanics the American system which replaced the wooden drill rods with manila rope. It was a significantly faster method but one which required greater skill on the part of the operator with success dependent on a driller's correct judgment and on his regulating contact between bit and bottom. Also it was not especially well-suited to the geological conditions of the area. By the time the Americans left—they did not care much for life in Baku and lasted only a few months—the Nobel engineers were experimenting with a modified rod system, utilizing circulating water rather than a standard sand-pump to clean out the bore hole. Hollow small-diameter piping was used in place of the rods, and as drilling proceeded water was periodically—about every two feet—forced down the pipe by a steam engine which generated a pressure of fifty pounds per square inch, sufficient to push the debris up the sides of the casing. This became the standard Nobel drilling system and in the first quarter-century of the company's history—up to the time the rotary method was becoming popular in the United States—it was used to drill more than five hundred wells at a combined depth exceeding a hundred miles, producing some twenty million tons of oil.[3]

About the time Bary was dispatched to Scotland and the United States, a Swedish engineer, Alfred Törnquist, was sent to America to learn all he could about the refining processes. It took him three years but when he finished he was the beneficiary of all that the Americans had discovered and devised in the 1860s and 1870s about the new science of refining: the trials and errors of still construction, the proper placement of piping, the selection of the proper metal to withstand the intensive heat of the furnaces, the use of seamless wrought-iron and seamless steel bottoms properly curved to minimize the contraction and expansion of joints connected to the sides of the still, the design of condensers, the practicality of steam distillation. He studied the vertical, cylindrical cheesebox stills, the most popular type then in use in the United States and reputed to render a higher-quality kerosene at less cost. But opponents of the cheesebox criticized its high maintenance costs and the longer hours of operation needed for refining. The average size of the American stills was between five hundred and six hundred barrels, although there were some which exceeded three thousand and giants at thirty-five hundred barrels. The newest American refineries seen by Törnquist, however, did not exceed a thousand-barrel capacity.

When the Nobel engineer returned to Russia he and Ludwig made immediate plans to change their basic method of refining, to alter radically—over Robert's strong protests—the refinery opened in 1875. With the imposition of the import tax on petroleum products and the subsequent opening of the Russian market, capacity had to be increased substantially to take advantage of the opportunity; but Ludwig also wanted to introduce a new system, one based on continuous distillation. The other Russian refiners did not copy the system and it was not used in the United States for another twenty-five years, yet it was unquestionably the most efficient method of refining and it enabled Nobel to refine a higher grade of kerosene and other products at less cost than the standard single-still system then in use throughout the world. It also made Nobel the largest refiner in Russia. Mirzoiev, Sarkissov, and the Baku Oil Company were the big three of the Baku refiners when Ludwig's new distillery was brought on stream in 1882. Three years later Nobel .was the undisputed leader, refining more than four times as much as the second largest and more than the next five producers combined.[4]

As early as 1860 in the United States, the year after Colonel Drake brought in that first American well in Titusville, a patent was granted for a semicontinuous distillation still for refining coal oil; in 1867 and 1870 two other patents were granted for similar refining of petroleum. After fifteen years' work Samuel van Syckel was awarded a patent in 1877 on his system utilizing a battery of three shell stills; and about the same time a Baku refiner, Tavrizov, applied for a patent on a continuous process which was more or less a modified alcohol-rectification system. It was not until 1881 when Ludwig and Törnquist installed their new system in Baku that the world had the first multistill continuous system ever adopted for commercial use.

From the field the oil moved by pipeline to huge reservoirs dug into the ground with walls of earth ten to fifteen feet high, clay-lined, then covered with planks resting on wooden scaffolding and coated with asphalt to prevent evaporation and percolation (most of the other refiners left their reservoirs uncovered). From reservoir the crude was pumped into a series of sixteen hundred-ton tanks. Whatever sand remained settled to the bottom before the oil was pumped into another series of brick-lined tanks through which pipes were laced to carry the heated residue being discharged from the stills at the other end of the process when the refining was completed. These pipes heated the crude to 100 degrees Centigrade and it was then pumped into a third series of tanks. The temperature was doubled before the oil entered the first of sixteen stills in which it was heated to progressively higher temperatures until it left the last of the stills between 300 and 320 degrees Centigrade. The battery of stills was arranged in a manner which gave the name to the process: each successive still with its higher temperature was placed lower than the preceding one so that the oil flowed by gravity continuously through the entire series of stills.

The vapors from the heated oil were piped into air and water condensers and the distillates into what was termed a tailhouse where they flowed into an open tank divided into several compartments. Distillates from the various stills reached the tailhouse separately but they were all combined into one of three products: kerosene, gasoline, and the distillate from the air condenser which was called black kerosene. This was redistilled and divided into a light and a heavy product; the lighter one went to the benzine refinery and the darker to the kerosene refinery where it was added to the kerosene. The gasoline product from this initial stage of refining had the lowest specific gravity and was then distilled by steam heat and further separated into a light and a heavy distillate, the lighter proceeding to the benzine refinery and the darker to the kerosene refinery where it too was added to the finished product.

Ludwig's new system not only produced the highest quality kerosene in the country through its superior design, solid construction, and technical supervision—samples were constantly checked to ensure uniformity; it produced that kerosene more economically and with greater efficiency and speed than the single-still process used by the other producers. Thirty-five percent of the crude was made to yield kerosene; some of the other refiners producing that Baku Sludge had yields as low as 20 percent. Also there were no interruptions in the refining process for repair and the necessary periodic draining and cleaning of stills. Nor was there delay from preheating of the crude. The flow and refining was continuous; one or two stills taken out of use did not shut down the whole operation.

Postrefining was similarly controlled for quality and consistency. The kerosene was pumped into agitator tanks arranged in pairs. The first tank was lined with lead to prevent corrosion of the iron by sulphuric acid which was mixed in a proportion of .694 per 100 parts kerosene. The mixing, by blasts of air forced into the tank, lasted six hours, and when completed the acid settled and was drawn off at the conical-shaped base of the tank. The second pair of tanks contained caustic soda in a solution of .273 per 100 parts of kerosene. This too was mixed for six hours and the settled soda drawn off from the bottom of the tank. This double cleaning process then was followed by a washing with water before the distillate entered one of the iron storage tanks on its way to the Nobel distribution network. Color and flashpoint were at the desired level, guaranteed by the careful control of mixtures of the different distillates as determined by the frequent laboratory testing of samples. In the American refineries it was often necessary to expose the distillate to sunlight or to spray the product in a color-lightening aeration process. The Nobel refinery relied strictly on quality control of the refining process.

The idea of using sulphuric acid as a kerosene purifier did not originate with the Nobels. A German chemist, Herman Frasch, had discovered the process while working in the United States and the great Russian scientist, Dmitry Mendeleyev, learned about it while on an American visit. The Petersburg professor had been commissioned by the government in 1876 to study both the Baku and Pennsylvania fields, to report on industry potential, and to make scientific comparisons. Robert was the first to take full advantage of Frasch's discovery, the first to use it properly in Baku, and the first to use caustic soda as a cleaning agent. Both methods were adapted for use in the new continuous distillation refinery. When that was brought on stream Ludwig also put into operation a new refinery, one for lubricating oils. He and Robert had bought a small installation from the Pertzov Company, rebuilt and expanded it into a nine-still twenty-four-ton unit. But their engineer, Hjalmar Crusell, wanted still greater capacity and in 1884 he installed a monster hundred forty-two-ton still. The workmen quickly christened it "Ivan the Terrible." Producing more than twenty-seven thousand tons a year Ivan proved worthy of the name when it exploded after ten years in operation. It was one of the few accidents that ever befell the Nobel refineries.

As the quality of refined oil steadily increased Ludwig and his engineers, not content to rest on past achievements or to remain satisfied with present profits or progress, decided to cut costs of refining still further by building their own factories for production of sulphuric acid and caustic soda. The acid plant was completed first in 1884 and used Persian sulphur mined just over the border; in a few years it was supplying almost a third of Baku's total consumption. The new soda-regenerating plant was finished two years later and provided all company needs as well as those of other refiners. In ten years annual production reached seven hundred tons and by 1907 over a thousand tons.

Ludwig also explored the feasibility of refining gasoline, then useful in the manufacture of rubber. With Baku crude yielding only 0.5 percent gasoline and the internal combustion engine still a few years off there was little need for great quantities, but Ludwig went ahead with his plans. The first gasoline refinery in Russia flew the Nobel flag and in the years prior to World War I that refinery became increasingly important. Nobel laboratories established by Ludwig researched ways and means of increasing the yield and purifying their end product. In Ludwig's time the laboratory worked mainly on kerosene, lubricants, and fuel oil.

The idea of using black gold to fuel the engines of industry and transport was not a new one; as early as 1858 at the temple of Surakhany the Transcaspian Trading Company fired their boilers with the natural gas jetting from the earth in those eternal pillars of the fire-worshipers. A few years later the Russian navy was experimenting on the Caspian with the use of petroleum bricks composed of oil and pitch. The Americans and British were seeking similar substitutes for coal and the British scientist Aydon along with the Russian Spakovsky, each working independently, deserve the credit for finding a way to pulverize raw oil and to blow it into a furnace in the form of a spray. Spakovsky's invention was installed in the first oil-burning steamer on the Caspian, the Lebed Company's *Iran* launched in May 1870; they then added other oil-burners as did the Caucasus and Mercury Steamship Company. By the end of the decade the navy had oil-burners on their eight ships stationed in the Caspian. It was the first oil-burning fleet in the world.[5]

Two Swedish engineers of the Caucasus and Mercury Company, Benkston and Sandgren, invented improved variations of the Spakovsky burner and a Baku engineer named Brandt devised an adaptation for use on the Transcaspian railroad. His device consisted of an overflow, trickle, and spray system in which the oil entering through a central supply pipe overflowed onto a small diaphragm, trickling down to a lip holder where it was driven into the furnace by a steam spray. It was a good system but Ludwig improved on it by adding a conical head with spiral grooves to give the entering flame of oil a rolling motion, thus sweeping it along the inner surface of the boiler.

He used a variation of this for stationary boilers: an arrangement of shallow troughs superimposed one over the other at the door of the furnace in such a manner that the fuel fed into the top trough overflowed into the one immediately below it down to the last trough. There was sufficient separation between the troughs to permit a current of air to sweep over the surface of the burning oil and carry the flame into the furnace. This method, Ludwig demonstrated, was extremely efficient: a pound of fuel oil burned with the overflowing trough system evaporated fourteen and a half pounds of water in a boiler; the spray system evaporated twelve pounds, while coal only vaporized between seven and eight pounds. Similar savings were realized from Ludwig's invention for locomotive and ship boilers. He could prove that his method used only five to seven pounds of fuel an hour while Brandt's and the others consumed as much as eleven to seventeen pounds an hour.[6]

Ludwig was trying to improve and rationalize where others, blinded by the supplies of available oil, were content to use the old methods regardless of waste. And those methods were still less expensive than coal or wood. Russian fuel oil produced almost twice as much energy as coal, and when that coal had to come all the way from England the economic savings and trade balance savings were obviously considerable. Baku, for all its remoteness, was certainly a closer and more reliable source than far-off Newcastle and the very nature of the Baku crude encouraged its use. Not only was it spouting from the earth; between 60 and 70 percent of it yielded residual oil suitable for fuel, compared with less than 20 percent in the Pennsylvania product with its higher percentages of the more volatile hydrocarbons.

By the 1880s the Russian petroleum industry was already becoming a fuel-oil industry, one in which the crude—the fuel oil—was more important than the finished and refined product. At the same time in the United States, with its centralized control of refining by Rockefeller and his Standard Oil Company, there was almost no use of fuel oil. Kerosene was the primary product and it was not until the opening of the Lima, Ohio, fields in 1885 that there was a ready supply of crude. When that crude was pumped to Chicago by pipeline the American transition to fuel-oil usage really began. For Standard the incentive was in the very nature of the crude. High in sulphur, it proved impossible to refine out the disagreeable odor. Standard, ever able to adjust to difficult marketing situations and unable to sell it as kerosene to compete with its Pennsylvania product, formed an "Oil Fuel Brigade," a team of technicians and salesmen who beat the hustings in the midwest to convince businessmen, hotel, factory, and warehouse owners of the advantages of fuel oil over coal, to convert coal furnaces to oil burners, to distribute free samples. Fast-growing Chicago soon led the nation in fuel-oil consumption, continuing that lead even after Standard was able to remove the sulphur and to refine a better grade of kerosene at its new Whiting, Indiana, refinery.

In the western United States Leland Stanford's Central Pacific Railroad experimented with the use of fuel oil in 1885 but it was several more years before the Southern Pacific put on the first regular oil-burning locomotives in the country. In Russia it was initially the railroads which consumed most of the nation's fuel oil but with industrialization the use soon spread to the metallurgical industry and to factories in the north. By 1890 the annual rate of consumption exceeded two hundred ninety thousand tons.

After Ludwig's development of new oil burners no further Russian improvements were made; when more efficient American and British units came on the market the Baku producers and shippers usually ignored them. They were too conservative to adopt still another new burner, or unable to circumvent the import tariffs, or perhaps just too sluggish from their own prodigal approach to the overabundance of easily-produced oil. Then too the level of technical skill, the training of the labor force, meant that only a minority of engineers and machinists would keep the furnaces clean, the burners in optimum working condition. Valves and pipes were ignored, furnaces allowed to run down to a state of disrepair and neglect when replacement was the only alternative.

Nobel burners, like their ships and their refineries, were the exception; but the Nobel organization had an infinitely stronger control of both labor and its routine of work and it was able to command a higher discipline and loyalty because of its concern for the workers' welfare. Ludwig's adamant insistence on improving the lot of the worker, on providing decent housing, a living wage, safe and secure conditions, recreational facilities, and opportunities for self-improvement had previously been demonstrated in the wilderness of Izhevsk and in the factory complex on the Sampsonievsky. In the sands of Baku—in the dusty, grimy, backward frontier dominated by greed in an atmosphere of Byzantine "baksheesh" and Oriental intrigue—he again demonstrated his enlightened philosophy.

In addition to his profit-sharing plan—after a distribution of 8 percent of the profits to shareholders, 40 percent of the remainder was given to salaried personnel and the remaining 60 percent to shareholders—Ludwig provided improved housing for the wage-earners far superior to anything previously seen in Baku. He offered free education, established technical schools in Black Town as well as in Balakhany, and ordered construction of what became famous throughout the Caucasus and the international oil community: Villa Petrolea, a compound near White Town, a walled-in oasis of homes and apartments for the executives of the company resident in Baku. Soil was brought in from Lenkoran a hundred miles down the Caspian coast, trees and shrubbery were imported and kept alive by the waters of the Volga brought back in the empty tankships. Administrative and technical chiefs, engineers, and various section and departmental heads lived in the villa furnished with a library, billiard and meeting rooms. There were gardeners, watchmen, drivers, and in 1880 the first telephone in Baku—a Bell apparatus connected to harbor and oil fields. The twenty-two-acre compound provided a model for later adoption by the industry: the care, feeding and recreation of oil men in the scattered outposts of the world. To the later generations of engineers and executives Ludwig's Villa Petrolea was probably as great a contribution as the tankers, the distribution network, and other innovations that poured forth from the fertile mind.

For the workers in the fields colonies of houses were erected, including separate quarters for the married men. At a time when most laborers in the oversized Russian enterprises were crowded together in miserable barracks without regard to sex or married status the Nobel workers had separate accommodations far better than anything they could find in the town. Ludwig refused to agree with his fellow capitalists that squalor was the workers' natural state, and in providing for his employees' welfare he was as unusual in Russia as he was in most other countries of the world. Typical of the prevailing sentiment was the fate of the Minister of Finance; he had recently been driven from office as a dangerous radical, condemned for advocating child labor laws and for championing a few abortive efforts to correct some of the worst abuses of the time. The other Baku producers, with few exceptions, had no interest in their labor force. Their sole concern was the price of oil and the number of rubles in their own pockets.

Among this group Ludwig's reforms and concern were most unpopular. But Ludwig was accustomed to making his own decisions, travelling his own road; and he was certainly used to opposition, direct and indirect, silent and vocal. But no matter how violent or bitter or stubborn that opposition, Ludwig obstinately and optimistically charged ahead. He led the way and showed those around him how to overcome. "That they should have never been discouraged by the opposition they met at every step they took," wrote British historian Charles Marvin in 1884, "is a remarkable testimony to the unflinching courage and irrepressible perseverance of Ludwig Nobel."[7]

CHAPTER SIX

BRANOBEL and the Brothers Nobel

Robert's little refinery had turned into a gigantic enterprise practically overnight. His 25,000-ruble investment of "walnut money" was transformed into a multimillion-ruble operation with a distribution network spreading to every rail line and waterway in the country. But it was Ludwig who was responsible, Ludwig who implemented the grandiose schemes, Ludwig who did the designing and planning and paying. Somewhere along the line Robert, who started it all, was lost in the hectic happenings generated by his brother's feverish progress.

Never one to surrender control easily and understandably jealous of his squatter's rights as eldest brother and pioneer in the area and the industry, Robert planted himself squarely across the trail Ludwig was blazing. A frequently choleric temperament and occasional inability to coexist with fellow workers complicated the issue, embittering his reaction to Ludwig's takeover. He recognized that Ludwig would continue to dominate and that both his brothers, with their far greater financial resources, would certainly control future plans and developments. Thus he opposed formation of a shareholding company and from 1879—once that company was established—he opposed most of what Ludwig was trying to do to organize an industry. He particularly resented the plans for continuous distillation. Ludwig's insistence on this improvement might have been what finally convinced him he had to leave Baku.

In the fall of 1879 Robert suddenly departed, telling no one his destination. Eventually he informed his family that he was in Switzerland for a rest cure, and he did return the following spring but remained only a few months: the Swiss air had not improved his temperament or his attitude. Ludwig confided to Alfred that their brother had become impossible, that his actions were creating serious unrest and many problems among the workmen in Baku. Despite the fact that everything was going so well with increased production and what Ludwig described as "progress on all fronts," Ludwig was receiving "hateful" and "bitter" letters from Robert. He could neither understand nor do anything about the sad, strange behavior.[1]

Sickness, usually consumption, was given as the reason for Robert's leaving—the polite explanation to save embarrassment. To be sure, eight years in Baku could wreck the constitution of many healthier men, and there was some truth in Alfred's comment on learning of the departure that Robert could now "guard his health from morning to night."[2] At heart, however, Robert was not the kind of individual to take orders from a board of directors who were late arrivals on his particular scene. He was far too jealous of his own independence, his own freedom, and was as impatient with the mistakes of others as he was with their instructions. He was probably not too eager to leave the stage of his greatest triumphs, but at age fifty-two he had little desire to continue to be merely the manager of the Nobel Company in Baku and even less desire to return to what he called "the swamp of St. Petersburg." As Dostoevsky observed, "the mere influences of climate mean so much," and Robert no doubt agreed with the judgment in *Crime and Punishment* that "there are few places where there are so many gloomy, strong and queer influences on the soul of man as in Petersburg."[3]

The only alternative was retirement in Sweden. In return for a generous cash settlement and a percentage of shares in the company he surrendered all rights and responsibilities and moved to a comfortable estate on the Bråvik Fjord. "Better freedom and herring in Sweden than riches and servitude in Russia," ran an old Swedish folksaying. As the eldest of the three surviving sons Robert had already spent more time in Sweden than either of his brothers, had received five years of Swedish schooling compared with Ludwig's three and Alfred's single year, and he now returned to the land of his birth, there to live until his death in 1896. He was neither more nor less Swedish than the other two; each of the Nobels represented to a greater or lesser degree—in personality, temperament, lifestyle, and general career—manifestations of what those who observe and record such things regard as the Swedish character.

International in outlook, lacking narrow nationalism or provincial patriotism, the Nobels—like so many other Swedes from the time of the first Viking colonizers—were able to adapt easily to foreign cultures. Whatever truth there is in the saying, "Scratch a Swede and you will find a Russian," there can be no doubt that the thousands of Swedes who settled in all areas of Russia through the centuries melted into their adopted society without retaining for long that sense of separateness which charac- terized the German enclaves or, elsewhere in the world, settlements of French and Italians. Self-sufficient, independent, with a devotion to work and a reliance on its routine, the Swedes demand a high level of order in that work, a sense of purpose and perspective in their own rational slide- rule civilization. For the Nobels, their boyhood of poverty strengthened the motivation. "He who does not work need not eat," declared Ludwig.

Immanuel believed that it was Alfred who demonstrated the greatest in- dustry in youth and Ludwig the greatest genius, but when comparing achievements it is indeed difficult to assign any priority. In common with their fellow entrepreneurs in the United States, born like Ludwig and Alfred in the decade of the 1830s—Andrew Carnegie, Hill, Harriman, Jay Gould, Wanamaker, Mark Hanna, Marshall Field, Rockefeller and his partner Flagler, Pullman, the elder Morgan—the two Nobels displayed in equal measure dazzling talents of genius and industry, the vigor, imagina- tion, initiative, and inventiveness which typified the great captains of in- dustry and trade. Powerful wills and indomitable energy, an obstinacy often as unyielding as the iron in the rocky Swedish soil—equally essential qualities for those who dared create new industries, giving life across vast virgin territories to their visions and dreams.

For Alfred, execution did not mean direct and total personal involve- ment and responsibility. His rule was "never to do myself what another could do better, or at any rate, as well." In contrast to both his brothers, he was certain that "if you try to do everything yourself in a very large con- cern, the result will be that nothing will be done properly," and whoever tries to do it all himself will be "worn out in body and soul and probably ruined as well."[4] And ruin, the threat of financial ruin, was never far from the nightmares of Alfred who was almost too Swedish, too northern, in his acute awareness of impending tragedy, his sense of doom, his fears that someday he would be forced to the wall and bankrupted. His father had been there twice; the sons needed no reminders of the agonies and frustrations that follow a fall from the heights. And if the memory should weaken, they were surrounded by examples of broken men and there was a popular contemporary play by Bjørnstjerne Björnson about the fall and bankruptcy of a Norwegian tycoon.

Robert did not seem as morose about what fate might bring, although he appeared to have been the most influenced by that childhood of want, the experience of peddling matches. Seldom one to delegate authority, frequently mired in petty details and often petty squabbles with associates at all levels, he lacked the tenacity of either of his brothers and certainly never revealed the ebullience of Ludwig, that extrovert of restless mind, unable to resist the challenge and joys of personal involvement, the thrills of devising some new method of production, some new machinery, of planning vast organizations, promoting the welfare of his workers. Ludwig was not a person easily forgotten, even after the briefest encounter. He was, in the words of one of his closest collaborators, "a personality in the fullest sense of the word." Short in stature but broad-shouldered, with bushy eyebrows shading blue-gray eyes that had an almost hypnotic quality, he dominated any audience, revealing at once an inner clarity of purpose and deep concentrations of power.[5] Unlike Alfred who lived and worked aloof from his employees, Ludwig's home was in front of his factory, his office a room of his house, and he spent many hours with his engineers and draftsmen, his factory foremen and section chiefs. It showed: not only in his own frequently brusque, direct manner, but also in the explicit loyalty of his workers.

Ludwig's philosophy of business reflected a basic difference between the brothers; unlike Alfred, he believed in no secrets, no monopolies, no special privileges. The distinction was apparent in their personalities, with both Alfred and Robert often revealing that other side of the sometimes merry Swedish exterior, that "silent touch of the somber and hidden sadness of the pine forests," as a Danish professor observed many years ago.[6] The silence and solitude of the dark forests, the yearning for isolation, the overwhelming melancholy—all so brilliantly portrayed in the films of Ingmar Bergman. His professor in "Wild Strawberries," condemned to loneliness by the Inquisitor, could have been patterned on Alfred the recluse who lived with the fear that he would die away from family and friends, alone in some strange setting, comforted only by servants.

Like Robert, Alfred could disappear for long periods of time without informing anyone. For Ludwig with his large family—eleven of his eighteen children survived infancy—and endless responsibilities as leader of two dynamic enterprises, such escape was as impossible as it was unnecessary. Despite his practical talents, his administrative and technical skills, he was not a typical Swede with that lack of interest in people, that characteristic cited for decades as the reason why every second Swede is an engineer, why they have such interest in machinery, why they are fascinated with what can be touched and felt. Ludwig added a genuine social consciousness, a sincere interest in the well-being of those on his payroll. At heart an irrepressible optimist, he probably was as hopeful about mankind as he was about his own capabilities and career. This again put him at odds with both brothers who shared a more pessimistic, cynical view of their fellow man. In Robert that cynicism could turn into jealousy and suspicion; in Alfred it helped mold his radical political and religious sentiments.

All three brothers read in French, German, English, in addition to Russian and Swedish, and Ludwig was reported to favor Voltaire above all authors; but it was Alfred the bachelor who was the bibliophile, who devoured the works of those two other part-time expatriates, Ibsen and Strindberg, and it was Alfred who, from all external evidence, was most challenged by the complexities of the intellect. His socialist political proclivities, his excellent library, his effort at authorship: these were as remarkable manifestations of the genes of Rudbeck genius as those of Immanuel's genius were revealed in the yacht he once designed to be built of aluminum.

A not so fortunate inheritance shared by all the brothers was that of delicate health. But Ludwig and Alfred pushed themselves at such a pace that physical exhaustion probably would have overcome far hardier types. The rigors of Petersburg winters and Baku summers were hardly conducive to a preservation of that health, and despite all other problems the combination certainly took the measure of Robert, sending him home to the shores of Sweden.

Swedish economist and Harvard professor Eli Heccksher once praised Ludwig as one who "in terms of breadth of interest, leadership qualities and general characteristics, was perhaps at the forefront in this richly gifted family";[7] but without Roberts's alert recognition of the immediate potential of Baku, his intrepid determination to commence development, Ludwig might not have had the opportunities to widen his entrepreneurial horizons and devote his talents and energies to a new industry. Or his fortune.

The numerous projects and innovations initiated after his first visit to Baku clearly were going to cost a great deal of money. Some provided self-bailing potential—the pipeline paid for itself in its first year and the *Zoroaster* in its first season—but the initial investment sums had to be found whether or not resultant savings in freight costs could be transferred quickly to new expenditures. When the expansion was constant, the demands of Ludwig's new infrastructure never ending and always increasing, those sums were not always available.

Robert's new refinery did show a profit, but his start-up cost had been 450,000 rubles. A third of that amount was advanced by Ludwig's Finnish mining partner and longtime friend Baron Standertskjöld. He extended the same kind of support he had given in the early 1870s—the use of his securities as collateral for bank loans; he was also instrumental in arranging loans from abroad through several Finnish banking houses. Ludwig's other comrade-in-arms from Izhevsk, Peter Bilderling, joined forces in 1877, putting up 300,000 rubles. "If the word gentleman had not been invented," Alfred once declared when discussing this Russian officer, "it would have to be to describe Bilderling and his brother Alexander." Alfred himself invested 115,000 rubles, and as family financial expert advised Ludwig that he should go public. A shareholding company was the only way to raise the kind of money Ludwig was spending and obviously would continue to spend. In May 1879 that corporation was formed.

Tovarishchestvo Nephtanavo Proisvodtsva Bratiev Nobel (the Nobel Brothers Petroleum Production Company), BRANOBEL for cable communications, was soon accepted everywhere simply as Nobel. Ludwig, Robert, and Alfred were the main organizers along with Peter Bilderling. Six hundred shares at 5,000 rubles each were issued. Ludwig, who had purposely set the share price high to discourage market speculation, was the majority shareholder. Of the 3 million–ruble capitalization ($2.5 million), he had 1.61 million, Alfred, 115,000, Robert, 100,000. Bilderling held 930,000 and his brother Alexander, 50,000; I. J. Zabelskiy, a Petersburg businessman, 135,000; the chief accountant of the Petersburg factory Fritz Blumberg, 25,000; the factory's chief engineer Mikhail Beliamin, 25,000; A. S. Sundgren and Benno Wunderlich, 5,000 each.[8] Sundgren was a Swede, and Wunderlich a wealthy German in the Moscow import business with a magnificent castle in Dresden where he retired after serving for years as chief of the Nobel agency in Moscow.

Capitalization was raised the following year by another million rubles; in 1881, 2 million, 4 million the next year, and 5 million in 1884, reaching a total of 15 million—a fivefold increase in five years and a valid barometer of the tremendous expansion of the company. At Alfred's insistence and for the purpose of stimulating the interest of the small investor and raising additional money, the 5 million float in 1884 was in 250-ruble shares.

When Robert left Baku in 1880 he was replaced by a Swedish engineer, Ullner. A few months later Törnquist took over; then Gustav A. Törnudd who served until 1888 and was by all contemporary accounts extremely capable, supervising field, factory, and fleet, ignoring nothing chemical or technical found anywhere else in the world and of potential use to Nobel, as a British visitor observed, praising the Swede as "an engineer of comprehensive mind as well as practical skill."[9] Baku was clearly in competent hands, and after Robert departed Ludwig did not have to worry about personality conflicts in the south.

Westward, however, from Paris, his other brother was beginning to cause problems. Alfred, by now a recognized international financial wizard, was worried about the corporation's expenses. He had considerably more experience than Ludwig in dealing with banks and bankers, in raising money, arranging credits, floating loans, trading on the Paris bourse, dealing in London money markets, negotiating in the boardrooms of the German banks. The expertise was invaluable to Ludwig; in the years ahead the reputation and financial solidity that supported such competence was even more valuable. But the tremendous capital demands to subsidize Ludwig's imagination and creations across the sands of Transcaucasia and the waters of the empire tried sorely Alfred's pocketbook as well as his patience.

Alfred's original participation in the company, 115,000 rubles, was relatively small; but as the company expanded and expenses doubled and then tripled, he was the only board member with the ready cash and credit to help finance that expansion. Chief accountant Blumberg kept him informed on all financial developments, on the status of bank loans, on plans for additional installations and capital requirements. As he read the reports his concern increased. The expenditures were endless, the rate of growth hectic. Ludwig was building and buying like a man possessed, a man who feared his time was running out. Just a year after the organization of the company the *Zoroaster* was in service and other tankers were already ordered, the storage and distribution center at Tsaritsyn was completed and similar installations in Moscow, Petersburg, and Riga were under construction. Large barges were being built and the first hundred rail tankcars were on the way. A large-scale and expensive refinery was being laid out and Ludwig was talking about buying another refinery for lubricating oil and putting up other factories to make his own acid and caustic soda. He was even importing soil and water and trees and bushes and erecting a luxurious villa for the Baku manager while building special houses for the workers and talking about schools, libraries, hospitals.

Alfred was greatly troubled. He had none of his brother's robust optimism. "The main point," he wrote Ludwig, "is that you build first and then look around for the wherewithall."[10] Alfred was far more conservative; he wanted all the financial support lined up and signed up before any contracts were let. For a time he was the one who provided the support. In 1881 he advanced 656,000 rubles to pay Motala for additional tanker construction. But when he heard that the *Nordenskjöld* had exploded and the *Buddha* was badly damaged in one of those sudden Caspian storms he thought twice about his brother's brilliant scheme to transport oil in tankships. At the same time there was another fire, this one in Petersburg in the factory's pattern section. The world press reported it to be in Baku at the Nobel refinery. Investors and bankers across the continent were alarmed and Ludwig immediately went to the papers to explain the facts, but the investment community had their scare and Alfred, with that somber sense of foreboding, was worried that someday there might be a fire at the refinery.[11]

There were enough problems in Baku without adding fire to the list. Overproduction had driven the price of oil down to levels where Nobel profits were seriously affected; an outbreak of fever killed several key Nobel engineers and technicians. Alfred's list of worries lengthened and in February 1883, when he advanced another million rubles, he was concerned enough to discuss it personally with Ludwig in Petersburg. That was as close as he ever wanted to get to the source of the irritation. He did not want to visit Baku. "The waterless, dusty, oil-stained wilderness has no attraction for me."[12]

Alfred tried to convince Ludwig to speculate with company shares on the market as a means of raising money but Ludwig regarded speculation as a refuge for those who were too lazy to work. He told Alfred to "give up market speculation as a bad occupation and leave it to those who are not suited for really useful work."[13] He did, however, finally accept Alfred's suggestion to issue 250-ruble shares in order to stimulate small-investor interest on the markets of Europe. Alfred could view the problem only through the mind of a financier or a bookkeeper: he argued that Ludwig had moved too far too fast, that he had not established proper lines of credit. He reminded his brother of his own experience in building a dynamite business in a variety of countries with a variety of currencies, market, and credit arrangements. He went further. When Ludwig did not heed all his advice Alfred, in a pique of anger, attacked his brother and his management of the company. Verbally and by letter he questioned Ludwig's ability, criticized sarcastically the plans for expansion, expressed serious concern about future prospects. After the defection of his older brother, these attacks and complaints by the younger one were all the more frustrating and caused Ludwig considerable personal anguish.

Nonetheless, Alfred did provide most of a 2 million–ruble loan—a syndicate made up the remainder—and laid the groundwork for an additional million-franc loan from the Credit Lyonnais in Paris. Ludwig and his faithful partner Bilderling personally backed the loan, to be repaid in six months through the April-September sale of kerosene. One hundred twenty-nine thousand tons of kerosene were projected for that period and at 50 kopecks a pood there would easily be sufficient return to pay off the loan. This use of future petroleum production as loan collateral was an important precedent; in a few years even the Russian State Bank accepted the idea.[14]

A few months later Ludwig received the welcome news that the State Bank granted him a personal credit of 2 million rubles at 7.5 percent. Petersburg banker Gunzburg was trying to squeeze 15 to 18 percent out of him and the confidence shown by the State Bank in the head of Nobel Brothers, who was thus able to avoid the usurers, had a favorable effect on the other financial institutions in and out of Russia.

The year ended on another good note. With the aid of a friend who was a member of the Prussian parliament and the incentive of a significant shift in German foreign policy, an additional 5 million rubles was raised.[15] Bismarck had let it be known that the long winter of isolation for Russia was over, that it was time to warm up Russian relations with favorable financial considerations. The Berlin Discount Bank responded with a ten-year, ten-month 6-percent bond issue. Their announcement listed assets of the Nobel Company at 14.71 million rubles with more than a fifth located in Baku, 5.5 million in railroad tankcars and twenty-eight storage depots, and just over 4 million in the Nobel fleet—a dozen tankers on the Caspian and eleven barges working the Volga.

Despite the change in financial fortunes, the success in raising money, Alfred continued to nip at the management, complaining that the company lacked qualified financial experts. He acknowledged that Ludwig and his team were eminently qualified in all aspects of engineering and general administration, but bemoaned the fact that in Russia it always had been impossible to find trained economists or money managers. As his attacks grew sharper in tone, Ludwig took more and more time to respond with long and carefully reasoned replies. But Alfred would not be consoled and he continued to fire sarcastic salvos from the fastness of his Avenue Malakoff mansion. His attacks were more than complaints about overextended credit, about Ludwig's insatiable urge to expand. Basically, he did not really understand the industry. The viciousness of the competition, the wild fluctuations in the price of a product dictated by totally unpredictable forces, the tremendous infrastructure needed to move that product from well to wick, the impossibility of using the main transportation artery for half the year—these were alien considerations to the man whose own industrial empire was based on a product he alone controlled. He alone determined the quantity of dynamite to release on the market; he alone decided which plants in which country should charge what price. Competition was of little or no concern to him and he could run his business from a hotel room in any country he chose.

With Nobel Brothers forever dependent on Russian materials and transportation and markets Alfred feared there could never be a stable situation; and with the world market at the mercy of the power of Standard Oil and the Americans there could not even be a guaranteed profit. To his dying day Alfred was pessimistic about the future of Nobel Brothers and could only view with often overwhelming negativism the entire industry. Robert's continued carping from the security of Sweden did not help matters. Alfred was certain he would someday lose great sums of money in the Russian oil business, and he sent Ludwig a twenty-eight-page memorandum on the financial problems of the company, proposing solutions and implying that he would like to be more closely involved in the financial decisionmaking process.

But Ludwig was sure that such involvement would cause more problems than it would solve and he was confident the company would weather all storms, meet all challenges, and emerge stronger than ever. He refused to engage in personal attacks—he had no time for them—nor did he run from the challenges. He exulted in the thrill of challenge, convinced that the powers of mind and body were strengthened by conquering and overcoming. He never doubted that when the battles were over, the struggles at an end, it would be the Nobel banner flying from the victory mast. And he was just as certain that it would not be hoisted up there by the treasurer of the company or some accountant or a financial marvel of a brother sitting in far-off Paris, his desk crammed with notebooks of figures and his safe deposit boxes stuffed with stocks and bonds.

Ludwig was not a bookkeeper but an engineer, not an accountant but an entrepreneur who was never content to delegate, to merely invest, but who always insisted on becoming personally and totally committed. When his brother delivered a particularly scurrilous broadside just before the end of the crisis year 1883, Ludwig reminded him that "far out in front of businessmen and bookkeepers stand men with heart, a sense of honor and a strong sense of duty."[16]

CHAPTER SEVEN

Europe's Second Thirty-Years War

Had Ludwig been willing to take Alfred's advice he might have been able to forestall if not prevent the invasion of his domain by the powerful Rothschild interests. But the cost would have run into millions. Expenditure and exploitation of such an amount would have required approval of a government already concerned about the growing power of Nobel and the near monopoly of an increasingly important segment of the economy.

The immediate issue involved was the railroad running from Baku to Tiflis and across the Caucasus to the Black Sea port of Batum, taken from Turkey in 1878. The Russian oil producers Bunge and Palashkovsky formed a syndicate of investors, were granted a government concession, and commenced construction of a five-hundred-sixty-mile alternative to that Baku-Astrakhan-Volga route which Nobel pioneered and for the most part controlled. But construction was exceedingly slow and expenses excessive, far more than the backers could provide. They turned to other Baku producers, but overproduction had seriously deflated the price of crude and none of the other independents were able to risk additional capital investment. Nobel also was suffering from the drop in market price—in later years the company benefited during such declines by purchasing at depressed prices and holding the oil in its extensive storage facilities. It was totally committed, in fact overcommitted as Alfred never failed to point out. But Bunge and Palashkovsky were not really eager to involve Nobel and they turned abroad, finding an eager ally in the French branch of the Rothschild family.

For years the name Rothschild had been synonymous with European banking and also with railroads. Back in the 1830s Salomon of the Austrian house had built the continent's first major line, all sixty miles of it from Vienna to Galicia. In the same decade Beau James, the resident Rothschild in France, opened the Paris–Saint-Germain as well as the

Paris-Versailles line; ten years later he financed the Chemin de Fer du Nord. In the 1850s Beau James's son Alphonse, along with cousins Anselm in Vienna and Lionel in London, bought the Venetian-Lombard line and merged it with the Austrian Southern Railway. The French branch of this Jewish royal family then developed the Mediterranean Railway. Investment in a Caucasian line would be no novelty to the Rothschilds. Then, too, they were heavily engaged in mineral exploitation, in financing mines in. India, Burma, and South Africa. The French brothers Edmond and Alphonse were very active in the rapidly expanding kerosene trade and had financed a refinery at Fiume on the Adriatic. For them, completion of a rail line from Baku to Batum meant that Russian crude would eventually reach that refinery and Russian kerosene could flow to western markets in the immediate future. The lower-priced Russian products would permit the Rothschilds to create a trade independent of the Americans and Standard Oil.[1]

In exchange for numerous mortgages on refineries, wells, and transportation facilities the French House of Rothschild extended nearly $10 million for completion of the railroad. In addition they arranged credit to many small independent refiners, paying in advance for their kerosene. In a time of depressed prices many of the smaller independents thus were saved from going under, but the commitment also made them dependent on Rothschild who was applying the techniques of control and conquer perfected by Rockefeller in the hills of Pennsylvania.

Ludwig was powerless to prevent the entry of Europe's greatest financial force into the Russian industry. The Rothschilds, with fine-honed antennae sensing economic opportunity, realized the potential of the industry and seized upon it as surely and swiftly as they had financed emperors and prime ministers, railroads and diamond mines; not like a Nobel—an Alfred moving from laboratory to financial statement, or a Ludwig deeply engrossed in every planning phase, sketch, and blueprint. The Rothschild interest was strictly that of banker, investor, not business manager or draftsman. They dealt in the world's currencies, the stock markets of Europe, and had the experience of generations in the counting-houses of the continent.

Ludwig regarded their kind of speculation as a cardinal transgression of honest labor and reward. A clash of the two philosophies was inevitable and the confrontations came during the most cruel periods of Russian anti-Semitism. The Russians had not forgotten that it was a Rothschild in London who had floated the 16-million-pound loan that financed the

English campaign in the Crimean War, but the blackest of reactions did not rage until the assassination of Alexander II in which several Jews were implicated. By an 1882 Imperial Decree Jews were forbidden to buy or rent any more land in the empire.

The law had little effect on the Rothschilds, however, and the year after the Baku-Batum railroad was completed they organized the Société Commerciale et Industrielle de Naphte Caspienne et de la Mer Noire (Caspian and Black Sea Petroleum Company), always known by its Russian initials as Bnito and destined to be Nobel's major Russian competitor. When braced by the threats of Standard Oil, Nobel and Bnito sat at the same table; otherwise they clashed from one end of Russia to the other, from one country of the continent to another.

On British entry into the fray—competing inside Russia for a larger share of the market, buying up Armenian and Tatar companies to gain a foothold in what by the end of the century was recognized in England as a lucrative investment—they fanned the fires. But after years of hesitation and conservatism they were only following the advice of their most energetic and outspoken oil pamphleteer. Charles Marvin had been the first to attempt to awaken his countrymen to the vast potential of Baku and its black gold, imploring them in the mid-eighties to look to Russia: "A business into which the proverbially cautious Rothschilds are throwing themselves with vigor surely cannot be considered unsafe for Englishmen."[2]

It was ten years after those cautious Rothschilds first appeared on the scene that the British arrived and by that time Bnito was solidly established as the number two producer, refiner, and distributor. Rothschild money and Rothschild credit, dispensed from the offices on Rue Laffitte, created a bottomless well of sustenance and stimulation for an industry that was always starving. With Bnito setting the pace the old Roman naval base of Batum was transformed in a few years to a hectic, bustling port.

The year before Batum fell to the Russians, James Bryce, later to be British Ambassador to the United States, visited the sleepy little village on the Black Sea and found it flat, marshy, and malarial; but he predicted that if Russia ever acquired it from the Turks it would become a booming port city, the outlet for Transcaucasian trade, an active terminus for a railway from Tiflis. The author of *American Commonwealth* did not foresee the importance of the city to the oil trade, but in other respects his prediction was accurate.[3] By the time he reached Washington the little village had joined Baku as one of the most remarkable oil ports in the world. Philadelphia, San Francisco, Constantsa, Port Arthur, Balik-Papan— none of these could compare or be even remotely as fascinating.

Alongside the wharves were huge barrel factories, plants for fabrication of tin shipping cases, towering storage tanks, long docking facilities where small Greek sailing vessels lined up to take cargoes of the precious fluid to the same Mediterranean ports Greek sailors had been servicing since the time of Ulysses. British steamers, newly converted into tankers, waited for bulk cargoes to carry to Rothschild refineries at Fiume or Marseilles. Cases of kerosene moved out to all points of the compass. Holding two 4-gallon tin containers—the tinplate was imported from Wales and the wooden cases and tin containers fabricated in Batum—the cases were soon familiar sights on the waterfronts of the world. In time Rothschild had the largest of the case factories in Batum; their plant employed fourteen hundred and was capable of a daily turnout of thirty-six thousand tins packed in cases.

More than five thousand vessels transited Batum harbor in the course of a year: small sailers loading a few cases, coastal steamers taking on barrels, large tankers pumping out at the breakwater where racks of pipelines allowed four ships to load simultaneously. If a storm swept across the Black Sea the ships fled to deeper water rather than risk collision in the crowded port; as one visitor on such a storm-tossed day described the scene, "when the harbor is full of shipping, the port presents a perfect pandemonium of infuriated choppy seas and underground swells, of ships rolling, pitching, plunging and ranging at their moorings, parting hausers, wire ropes and cables, and colliding with other vessels."[4]

On gaining possession of Batum during the 1877 Russo-Turkish War, the Russians were forced by the Congress of Berlin to guarantee the town's status as a free port without fortification or other military presence. But by 1886 after Rothschild and the other producers had worked their transformation, the government decided it could no longer tolerate this humiliation and it unilaterally abrogated that portion of the treaty. Batum once again became a naval base, well fortified from attack by sea, with a fleet burning fuel oil which was wisely stored twenty miles distant from the harbor at Kobouletti in the interior. The harbor was dredged to a depth of thirty feet and additional piers constructed, but there seldom was sufficient docking space for all the ships and several always were waiting their turn in the outer harbor. New buildings were erected in the town, including a postal-telegraph center on Mariensky street, a Georgian

Catholic Church, and a Russian Orthodox cathedral. When the British arrived they formed their own congregation along with the inevitable yacht club and cricket field.

It was Britisher Alfred Suart, that rather bizarre sportsman who made and lost several fortunes in his eventful lifetime, who brought the first British ships to Batum. Following Ludwig's lead he had previously taken to Baku the small tanker *Paady*, using the Nobel system of ferrying it down the Volga in sections. From 1885 to 1889 he converted sixteen other small steamers to tankships to carry Russian oil from Batum across the Mediterranean and, in 1887, across the Atlantic to Philadelphia—his *Chigwall* was the first British tanker to dock in the United States. One or another of his vessels—*Petrolea, Tancarville, Wildflower, Bakuin*—were usually to be seen in the harbor loading Bnito cases and bulk oil for the markets of Western Europe.[5]

Rothschild money in Batum caused as great and rapid a transformation as Ludwig's entrepreneurial genius had generated in Baku and it was inevitable that the two forces sooner or later would collide. This was recognized by both camps and it was precisely for the announced purpose of avoiding open confrontation in the marketplace that the appointed representatives of Nobel and the Paris House of Rothschild sat down together at the conference table. The Nobel board of directors, especially Alfred, had no desire to enter into a struggle for survival with the mighty Rothschilds. From his Paris residence and with his financial experience he had too healthy a respect for their abilities and power to risk such wild thoughts, and he was greatly relieved to learn that company delegates Mikhail Beliamin and Ivar Lagerwall were planning to attend a May 1884 conference with Rothschild representatives in Paris.[6] The meetings were the first of many that would be held over the following years between Nobel and Rothschild, Nobel and Standard Oil, Rothschild and Standard, with other independents occasionally in attendance, reaching armistices and temporary truce treaties, breaking other agreements in order to crush former allies. It was another Thirty Years War and it was fought with rules of economic warfare which, relative to the time and tenor, were almost as uncivilized as those which characterized Europe's earlier endless and barbaric struggle.

Jules Aron, the able executive selected by Alphonse Rothschild to take charge of his new Russian interests, was chief representative of his firm at the first meeting; he briefed the two Nobel officials on the background of Rothschild interest in Russian petroleum, the initial proposal by

Alphonse's son-in-law to invest in the industry, the consequent formal and detailed study of all aspects of the idea, the final acceptance based on the potential involved, and the already heavy investment of the Paris house in the refineries in Austria, France, and Spain. It was an impressive presentation and Aron, who requested that the meeting be kept secret, concluded by saying that it also had been decided that the Rothschilds, despite their wealth and power, should not enter the market alone or in concert with Nobel competitors, but only in cooperation with Nobel. He thus proposed that some agreement for future coordination or amalgamation be considered.[7]

On digesting this news Ludwig instructed his negotiators to propose to Aron that Rothschild purchase a quarter-interest in the Nobel Company, but only after the base capital had been raised from 15 to 20 million rubles. The price of the shares would be determined by independent outside auditors appraising the total worth of the firm, adding to the share value an extra premium based on the current value in comparison to the value at the time the company was chartered. To protect the original shareholders he also proposed that, rather than dividends, a 6-percent interest be paid on the new shares until all the new facilities planned or under construction were in service.

Those facilities were primarily in Batum where Ludwig already was negotiating to purchase the property of Bunge and Palashkovsky and their Batum Petroleum and Trading Company for 1.5 million rubles. He intended to use another half-million to add storage and case fabrication plants. In order to move the oil from Baku to Batum Ludwig figured that still another million would be needed to purchase additional tankcars and almost as much to construct a pipeline over the Suram Pass, a real barrier on the newly opened Baku-Batum railroad. He also planned to build a new benzol factory for a million rubles more, and wanted to make a few improvements in the Baku refineries. All told, he was considering a total investment of 5 million rubles. Undoubtedly he realized that this was a conservative estimate, but he knew also that any overrun would be paid for by the Rothschild purchase of Nobel stock and whatever loan arrangements he could make by issuing new obligations on the increased capitalization of the company.[8]

Ludwig, always the optimist about the future of his company, had no fears or false hopes about negotiations with the powerful Rothschilds. It was Ludwig, after all, who was the leader of the industry. He owned the mantle of the "Oil King of Baku." In Russia his name was certainly as well known as that of Rothschild, and in Europe the name was gaining fame through tales of Ludwig's achievements and the successes of Alfred. Ludwig had nothing to fear from a Rothschild and he was confident that the rumors he had heard were true, that they really had no intention of investing heavily in the industry. The 5- or 10-million-ruble investment he was proposing was certainly not much to the Rothschilds, especially if that sum would guarantee them an orderly entry into the market.

Three weeks later Ludwig learned just how wrong he could be. Beliamin and Lagerwall returned to the conference table in Paris, but were scarcely seated when they were told that the Rothschilds had studied Ludwig's offer and had rejected it out of hand. Aron was not at the meeting—he reportedly had an illness in the family—and two members of the second team, Kornhauser from Vienna and Linger, chief of the Fiume refinery, took his place. It was Linger who proclaimed that Rothschild would never consider participating in any company without having majority control. As the Nobel representatives reeled from that declaration, Linger warned that the Rothschilds were too important and powerful in the financial world for anyone to consider competing with them.

The message seemed clear enough and Beliamin replied to the suddenly aggressive stance with a polite acknowledgment of the might of the Rothschilds but a firm refusal to turn over control of the Nobel Company. He reminded the pair that Nobel certainly did not fear competition, but that such competition would be concentrated outside of Russia because Nobel Brothers already dominated the domestic market.

Kornhauser then stated that he was instructed to tell the Russians that the Rothschilds did not find the present time propitious for dealing with petroleum affairs. Braced by that bit of bluntness Beliamin and Lagerwall concluded that there was no reason to remain at the conference table, and the next day they returned to Petersburg to report to Ludwig who assumed that the question was closed. He was making other arrangements to raise his 5 million rubles when Aron sent a telegram suggesting another meeting in September. No details were offered or agenda proposed, but Ludwig was amenable. He dispatched Lagerwall.

En route, the Nobel official spent a few weeks in Switzerland on holiday and, oddly enough, ran into Aron vacationing at Bex. During several days of discussion Lagerwall, in good faith and in hopes of reaching what he—like Alfred—regarded as a necessary alliance, provided Aron with a great deal of specific information about the company, its history, its present condition, its plans for the future. As a graduate economist and treasurer of Nobel Brothers, he was privy to such information and willingly briefed Aron, who repeatedly assured Lagerwall that he was strongly in favor of an agreement—leaning toward Ludwig's initial idea and ignoring for the moment the declarations of Linger and Kornhauser. Based on Aron's statements, by the time of the September meeting Ludwig and Lagerwall both were convinced that agreement was imminent, that the Paris house would sign on Ludwig's dotted line and purchase 20 or 25 percent of the company. The majority of the shareholders in Petersburg were in favor of the scheme. But three weeks of negotiations in Paris brought the agreement no closer to realization than it had been in June, although Aron confided that Baron Alphonse was studying the matter personally.

Just as the meetings concluded without result, the Nobel offices in Petersburg were visited by two oil engineer–diplomats, Boverton-Redwood from England and F. W. Lockwood of Standard Oil. They were en route to Baku, Lockwood on what he described as a secret reconnaissance mission and Boverton-Redwood—previously employed by Standard as England's outstanding petroleum expert to answer complaints about Standard's kerosene—to gather material for what would be the encyclopedia of the industry.[9] Their discussions with Ludwig were completed about the time *Bradstreet's* reported from its "reliable informants" that the transportation and production facilities of the Russian industry were so "wretchedly inadequate" that they posed no threat to American domination of the world market nor would they pose a threat in the future, given the nature of the tsarist bureaucracy.

Lockwood was not one of *Bradstreet's* "reliable informants," and he saw enough in Baku to confide in Ludwig that Nobel and Standard somehow should reach an agreement, but he requested that his comments be kept as secret as was his visit to Baku. His trip came just a year after a senior Standard executive, making the rounds in Europe, wrote Rockefeller that some agreement definitely should be made with Nobel—not to purchase the company because the Russian government would never allow it, but to acquire a substantial number of shares and retain Ludwig whose "shrewd ability, his knowledge of the Russian business, his high connections and his experience in dealing with the Czarist bureaucracy, make him invaluable."[10]

Lockwood, of course, was speaking out of the other side of his mouth to the Rothschilds, but that was part of the grand strategy. In Paris Lagerwall continued to hear Aron's protestations of enthusiasm for cooperation; but by November it was more than apparent that there would be no cooperation, no alliance, no purchase of Nobel shares. The company treasurer did not think that the sides were that far apart, but Ludwig had a more realistic reaction and he started making arrangements for a loan at the Berlin Discount Bank. Lagerwall again met with Aron in January and the following month the Rothschilds invited Ludwig to Paris, but he was far too busy and also was suffering from an attack of bronchitis. The Petersburg winters always were hard on him but for escape he only had time to go to Baku.

With at least two Rothschilds looking over his shoulder he had to make absolutely certain his company could withstand a series of sorties or an all-out offensive. He revamped his sales organization, negotiated new contracts with all domestic and foreign agents, gave them a monopoly in their territories, allowed them to buy at the lowest possible price and then set their own selling price. A new company director Gysser, member of several Russian bank boards and head of the Petersburg transport firm Schenmann and Spiegel—Lagerwall considered him to be the best acquisition the company had made since Törnudd—was put in charge of the sales department.

Abroad, a new sales network was organized in Austria under the Oesterrichische Naphtha-Import Gesellschaft (Austrian Petroleum Import Company), known as OENIG, with rights in southern Germany and Switzerland. The Wilhelm von Lindheim Company formerly held these rights. OENIG received its oil by tankcars loaded at the Warsaw depot and transported via Brody and Radzinilow to the border. Henri Rieth's company covered Belgium and Holland receiving its oil in barrels shipped from Petersburg, but he soon expanded his Antwerp installation to receive bulk oil and to serve as the central distribution point. In northern Germany the Berlin-based German Russian Oil Import Corporation, Naftaport, was strengthening its sales outlets, while in Scandinavia the local agents in Stockholm, Kristiania (Oslo), and Helsingfors were struggling to expand their insignificant markets. The firm of Bessler, Waechter and Company represented Nobel in England.

In France, between the power of the Rothschilds and the activities of the brothers Deutsch de la Meurthe—allied to the Rothschilds—there was no room for any other Frenchman, not to mention a Russian or a Swede. But Standard Oil managed to wedge in; in 1887 they negotiated a six-year contract with Deutsch calling for annual delivery of a half-million barrels of crude in exchange for Standard's agreement to keep its own kerosene out of France and to operate the Deutsch facility on the Delaware River.

Nobel did not have that much to offer. But it did have one French connection, a rather important one—the firm of Alexander André signed up in 1885 as Nobel's exclusive distributor of lubricating oils on the continent.

That contract and the new agreements with the other European distributors were all part of Ludwig's concerted effort to solidify existing sales arrangements and create new markets. This was also the time when he ordered the new Baltic tanker *Petrolea* at the Motala yards, personally paying the 165,000 rubles for the thousand-ton ship intended to carry bulk oil to Germany, Belgium, and Scandinavia. He also personally financed the new storage depot in Stockholm, but the company paid for similar facilities at Lübeck, Schweinemünde, Copenhagen, Antwerp.[11]

Eighteen eighty-five was also the year the converted steamer *Fergusons* was delivered to Batum and the 286-foot *Sviet*, another Motala ship, was put into service on the Black Sea. Owned by the subsidiary company Ludwig had organized for the burgeoning Black Sea trade, the Russian Steam Navigation and Trading Company of Odessa, the *Sviet* carried seventeen hundred tons of kerosene and was originally assigned to the Batum-England run. Had the demands of the domestic market not dominated company fortunes there would have been other Nobel ships on the oceans.

But even with increasing Russian consumption foreign sales were significant. From the first of January 1885 to the closing of the Volga in November the Nobels sold over sixteen thousand tons of oil to Austria, 85 percent of it kerosene and 15 percent crude. England was the second most important customer with just over eight thousand tons, followed by Germany with nearly seven thousand, Belgium with twenty-two hundred sixty, Sweden four hundred eighty, and half as much to Denmark. That same year close to a hundred sixty thousand tons were shipped on the Caspian-Volga route and the Baku reservoirs were still filled to capacity. There was every reason to believe that 1886 would bring still greater exports. Gysser and his sales staff were planning to sell more than fifty thousand tons in Europe with sixteen thousand tons each going to England and Austria.[12]

That was the plan. In late November Standard cut prices. Rockefeller's goliath already controlled more than 90 percent of all American oil exports and was the domineering force in all world markets except the Russian; but the price of monopoly is eternal aggression and when the Americans saw Nobel's sudden and successful invasion of their markets they quickly counterattacked. The effect on Nobel sales was immediate. The first offensive on the European continent, the first efforts to carve out new markets and seize new territory, came to an abrupt halt.

Not only did Standard cut prices. They resorted to sabotage and bribery. When one important customer was persuaded by the Nobel salesman to try the Nobel line of lubricating oils for his steam engines Standard bribed the chief engineer to increase the amount of water in the boilers. Water entered the engines and the engineer blamed the Nobel lubricants. But when the truth was learned the engineer was fired and Nobel became the sole supplier.[13] Standard no doubt fed its own version of the story into its own rumor mill. It was already circulating reports that the Nobel product was vastly inferior to Standard's kerosene, that it did not burn well and in many cases not at all. To consumers who had previously been exposed only to the Baku Sludge but not to the Nobel product the rumors had an impact and Nobel was forced to counterattack.

Naftaport's Berlin office distributed a flyer declaring that Nobel oil was at least as good as any other on the market, citing as authority the testimonies of the Royal Chemical-Technical Experimental Laboratory in Berlin, the "Kaiserlichen Normal-Aichungs Commission," and several chemistry professors from Karlsruhe and Antwerp. Naftaport invited comparable scientific analyses from any other individual or group and distributed a report compiled by a Bremen firm which had studied the German petroleum market during the previous two years. The conclusion? Standard Oil had done everything possible in order to eliminate, ruin, neutralize, or absorb competing Russian companies: price-cutting, circulation of rumors, anything that would remove the Russian threat and leave Standard in absolute control, in a monopolistic position to raise prices.[14]

But Standard was having its own problems in England. Customers were complaining about the quality of their kerosene; Lockwood was rushed to the scene and Boverton-Redwood hired to provide expert testimony. The first bulk shipments of Nobel kerosene were arriving in London and Standard officials were faced for the first time with a serious threat in a market they had long monopolized. They tightened their procedures for maintenance of barrels—the American kerosene was still being shipped in barrels—and reduced imports from the Lima fields. That malodorous Ohio oil would never keep customers from switching to Nobel's brand.

Standard was not about to let the British market slip away, but Nobel and then Rothschild were making important advances, were seizing some of the high ground. Their share of the British market increased from 2 percent in 1884 to almost 30 percent four years later when the Rothschild interests incorporated The Kerosene Company and the The Tank Storage and Carriage Company for the sale, storage, and distribution of their Russian products. Within a few months Standard organized its own British subsidiary, the Anglo-American Petroleum Company. The ink was scarcely dry on the articles of incorporation when orders were placed for ships, barges, tankcars, and reservoirs. Standard commissioned surveys of the major ports of England, Scotland, and Ireland to determine where bulk-handling installations could best be built.

It was reminiscent of the early days of Nobel Brothers. But the pricetag was considerably higher. As some indication of the rising costs of doing business in an industry already noted for its extravagant expenses, the same sum which had launched Nobel Brothers—$2.5 million—was now put into Anglo-American's new building program. Almost as much money went into Standard's German counterpart two years later. That company—the German-American Petroleum Society known as DAPG—was organized to perform the same function for the Low Countries and across the Rhine as Anglo-American in Britain.

Bessler Waechter, Anglo-American, OENIG, Naftaport, The Tank Storage and Carriage Company, DAPG—the opposing camps were in battle formation, the lines were drawn and the reenforcements in place. The greatest battles of the war were about to commence.

CHAPTER EIGHT

Exit the King

The rush by Nobel, Rothschild, and Standard to establish European bridgeheads in the widening theatres of the Thirty Years War was necessitated by the growing supplies of the refined product: the increased flow from Rockefeller's refineries, the opening of the Lima fields, the first trickles, then streams of oil coming into Batum on the newly completed rail line from Baku. To increase that flow it soon became apparent to the Russian producers and to the government—which took over management of the road as soon as it was finished—that the couple of hundred government tankcars shuttling back and forth between Baku and Batum were completely inadequate. The authorities therefore allowed those few companies owning rolling stock to use their own cars. Nobel immediately put some four hundred fifty into service, Bnito fewer than a hundred. To help make up the deficit and get more oil to their mushrooming installations in Batum Rothschild—ever ready with credit to exploit a given situation, to improve its position in the industry—lent money to the smaller producers to purchase tankcars. The producers had to agree to sell their refined oil in Batum to Rothschild's Bnito, but not at the Batum export price: they had to sell at the lower price prevailing in Baku.[1]

While this tactic guaranteed to Rothschild a steady supply of kerosene and continued capital domination of the Batum trade, it did not solve the basic barrier to continued expansion of that trade. To turn the stream into a flood more than a rail line and tankcars were needed. The obstacle of the three-thousand-foot mountain at Suram had to be eliminated. A tunnel was the best solution, but until that could be blasted out of the rock a pipeline was needed to carry the oil over a grade so steep it took two engines to snail just eight tankcars over the pass. Ludwig and his engineers had recognized the problem while the railroad was being built and had suggested a pipeline, but the government refused to grant them the concession to build it. Even after the trains were in operation and the severity of the obstacle apparent to all—mountain floods halted traffic for weeks at a time—nothing was done. But this time lack of action could not be blamed solely on the negativism of the authorities and the backwardness of the area. The government was pegging its rail-freight rates on the price of American kerosene, charging the producers the difference between the American price and the lower Baku price. Increasing the volume by piping Baku oil would affect this formula to the financial disadvantage of the government. The same reasoning held up construction of the pipeline over the entire length of the·railroad for another seventeen years.

But Ludwig kept up the pressure. With five of the largest Baku exporters he formed a syndicate—the first in Baku—to build a pipeline from Mikhailovo to Kvirili. He and his associates were interested only in sending kerosene through that line; they did not want crude being shipped out of Baku in tankcar or pipeline for refining elsewhere. Least of all did Ludwig—with the entire thrust of his operations across the Caspian and along the Volga, with his immense infrastructure in Baku—want to see crude oil pumped from Balakhany̆ to a Black Sea port with its own refineries for processing the oil or perhaps shipped in large tankers to refineries in other countries.[2] The smaller producers, the independents, of course wanted just that opportunity, that alternative of shipping out rather than selling to the giants during periods of overproduction. No less a personage than Mendeleyev called for the oil to be refined in Batum. But his voice and those of the independents were lost in the dry winds and when the government, seeing the success of Nobel's short pipeline over the pass, decided to lay pipe all the way from Baku to Batum it was strictly for kerosene, not for crude.

Nobel's forty-two-mile line was completed in 1889—four hundred tons of Alfred's dynamite were used in the process—but it was another seventeen years before the government finished that eight-inch $12-million line traversing the entire distance. The five-hundred-sixty-mile pipeline with its nineteen pumping stations was then the longest in the world, but it was a long time in construction. The government, never noted for its ability to expedite, stammered and stuttered across the Caucasus, reluctant to lose the control and revenue implicit in reliance on rail transport. It insisted that all the pipe and hydraulic equipment be of Russian make and threw up other obstacles. The inordinate delay prompted two Russian engineers with British backing to apply for a concession to construct a canal connecting the Caspian with the Black Sea.

With ever-increasing production from the Baku fields there had to be improvements in the means of transporting that production to the consumer, especially the consumer abroad. The Russian market had absorbed about as much kerosene as a poor economy could; the *mir* of the *muzhik*, the world of the peasant, did not include the luxury of regular illumination. Kerosene had to be exported. Ludwig, with his own well-organized distribution network, could move the kerosene all the way to Petersburg and the Baltic and beyond, but only Bnito could really compete in that area. The other companies had to ship what they could from Batum. If they did not they had to store and for most of the producers space was severely limited, especially in Batum where the trade for years suffered from a lack of tankcars in periods of great demand and lack of storage tanks in periods of slackened demand. Nobel was always there to purchase at the well and then to refine and/or store in Baku or ship to the storage depots in Tsaritsyn, Moscow, Riga, or anywhere else in the network. But even Nobel had limits and with the flow of oil almost doubling during the 1883–1885 period those new markets had to be found and better, faster means of delivery guaranteed.

The oil was pouring out of the sands of Baku in record quantities. The decade of the 1880s was even more remarkable than the 1870s, again proving to the world the truth of Mendeleyev's judgment that "the natural petroleum wells of Baku, as far as we are aware, have no parallel in the world."[3] But it was not until 1881 that Nobel hit its first gusher: well No. 25 at five hundred eighty-two feet threw a stream of sand two hundred feet into the air taking with it a ton of boring gear and requiring all the skill of the mechanics to cap and control the flow—the recorded pressure was two hundred pounds per square inch. For six straight months the derrick crew at No. 25 was drawing four thousand tons a day.

Eighteen eighty-three was a great year for gushers. Nobel's No. 9 hit at six hundred forty-two feet, spouting two hundred feet and spewing sand and oil over a two-hundred-foot radius, producing in its first month a hundred twenty-one thousand tons or close to thirty million gallons. Eighty thousand tons were immediately processed in Nobel refineries and the rest stored in reservoirs. Only four thousand tons were lost, despite the fact that the force was so great it prevented capping of the well. Such a small percentage of loss was a new record for Baku, an example of the value of adequate planning and training of the derrick crew. After six weeks of spouting, the engineers were able to build a platform over the well and force down with a battering ram a large mast to serve as cap; even then the flow was nearly sixty-four tons an hour. A few months later when Admiral Shestakov, Minister of the Marine, visited the area and expressed a desire to see a gusher, the Nobel drilling crew calmly removed the cap. There was a deep roar, then a sharp hissing sound followed by a two-hundred-foot column of oil. After ten minutes or so the admiral announced he was satisfied with the performance and the crew just as calmly recapped No. 9. In the year 1883 that kind of control in Baku was a remarkable achievement.

More typical was the eruption of Droozhba which came in at five hundred seventy-four feet with a pressure equal to Nobel's No. 9, or three times the norm. Tools were exploded three hundred feet into the sky, the derrick was completely destroyed, the ten-inch-diameter stream of oil and sand splattered over a hundred-yard radius; within fifty yards derricks were buried six to seven feet deep. It was like Iceland's Great Geyser. The rumble and roar could be heard for miles as eight thousand tons poured out each day. After one hundred fourteen days and two hundred twenty thousand tons of oil with two and a half times as much wasted Droozhba finally slowed down and was capped. But a few days later the cap blew off and for another three weeks the great gusher again wreaked havoc in the general area. An American engineer visiting Baku at the time was completely stunned by the sight; he said that in the United States Droozhba would have earned its owner a fortune, at least $25,000 a day. But for the little Armenian company which had just scraped together enough money to buy a tiny plot of land and erect a derrick the spouter spelled ruin. It could never pay for all the damages and was forced to bankruptcy.[4] Droozhba's assault on the marketplace also spelled disaster. The price of crude dropped to a quarter-kopeck a pood as thousands of tons of oil were burned off or allowed to run into the Caspian Sea.

Of the three hundred thirty-four wells in the Baku fields by the end of 1885 more than a third were producing; another forty were exhausted and shut down; fifty-seven were idled by damaged casing; another thirteen closed because of drilling accidents, and seventy-three were in the process of drilling with another nineteen ready to start operation. The following year there were eleven spouters, five from Nobel wells all of which were successfully capped with a combined flow limited to a thousand tons a day, then the full capacity of the Nobel refineries. Nobel had by far the largest of the hundred thirty-six refineries in Baku, but with the average daily yield of each active well running some thirty-two tons most of the refineries were working nearly continuously at full capacity.

Nobel's No. 15, hit at five hundred ninety-five feet, yielded nearly seventy-three million gallons; No. 18 only thirty million. Then came the fantastic fountains in the newly opened field of Bibi Eibat. One well alone gushed over eleven thousand tons a day, more than the total then flowing from all the twenty-five thousand wells in the United States, Galicia, Rumania, and Burma. Struck at seven hundred fourteen feet and streaming some two hundred twenty-five feet into the sky, the oil and sand were carried by a southern wind a mile and a half away, drenching houses, covering all of Bibi Eibat, creating a great lake which by the fifth day started flowing into the Caspian. On the sixth day when five or six thousand tons were roaring from the earth a strong breeze carried the oil all the way to White Town. The main square was blackened, the white rooftops darkened. On the eighth day there were futile efforts to divert the oil into old wells. For another week the oil shot into the sky and then slowed to a controllable rate of a thousand tons a day. In just over two weeks nearly two hundred thousand tons of Baku's black gold had been wasted.

The world had never seen anything like it. But that record was shattered the following year when the Mining Company's Balakhany gusher of three hundred fifty feet poured out sixty-five hundred tons a day for six weeks, burying derricks, pump sheds, roads, and houses. The company was unable to cap it and after six weeks the Nobel engineers were finally brought in (their gusher that year produced only a tenth as much oil), but only after a quarter million tons had been lost.

The area of production was only twelve square miles but the gushing oil from that small bit of earth made millionaires at an unprecedented rate. Among the most noted was the Armenian Alexander Mantashev. A six-foot three-inch giant of a peasant, he worked as a servant in the Baku office of one of the tsar's local military representatives. He saved some of his meager earnings and bought a small vineyard on the outskirts of town. Under the grapes were lakes of oil. In a few years he was one of three local Armenian multimillionaires and by the turn of the century a man of fabulous wealth and fabulous extravagances. The son, Leon, exceeded the father; he was Russia's greatest gambler, a collector of paintings, race-horses, and beautiful women. When he concluded a deal he personally took the gold rubles at well or reservoir or refinery, returning to his palace in town and throwing a party from the pages of the *Arabian Nights* with women from Eastern Europe, Oriental acrobats, gargantuan feasts and flowing wines. In the years before the first World War Mantashev swaggered through Europe like a modern Croesus, dispensing beneficence, keeping a large entourage, growing enormously fat with his movable feast and his permanent luxuries.[5]

In this primitive land a new style of Western dishonesty was combined with the ancient Eastern to produce a new and shrewder commerce which did not honor or even recognize the concepts of a later age. It was said that among the several hundred Baku oil barons, only ten were honest: an Armenian, a Swede, and eight Mohammedans. To quote one of the Mohammedans, Nobel was "the only one who despised Oriental methods of competition," always maintaining that "the European method was more successful anyway."[6] Ludwig was proud of his own particular guarantee and that of his company: "If you can find in Baku any man who can prove we are dishonest, cheat, adulterate or refuse to redress substantial grievances, we will face inquiry in your presence and if guilty, make amends."[7]

Russian expert Charles Marvin was enthusiastic in his admiration of the honesty and the achievement, informing his readers that "the brothers Nobel have acquired their wealth by honourable means and by enterprise and vision such as is uncommon even in the England of our time."[8] Or in the America of that time where Standard Oil was clawing its way to dominance by less than "honourable means"—private and illegal freight rebates, discriminatory price policies, bribery, and other secretive means by which the competition was suppressed.

The honest Armenian was Gregor Martinovich Arafelian who eventually became one of the wealthiest barons. Included in the Tatar total was Hadschi Abdin Zeinal Tagiev, surely one of the most interesting of all the oil entrepreneurs. He started as a hod carrier, worked up the ranks to mason and then stonecutter, and by the 1870s he was one of the most important building contractors in Baku. At the first auction of oil lands he joined with two Armenians to bid successfully for a small parcel in Bibi Eibat. After several years of frustration and no oil his partners withdrew, but Tagiev eventually brought in a gusher that yielded twelve thousand seven hundred tons. He built a refinery along the shore, added a dock, ordered his own tankers and then barges for the Volga. He established a sales office in Tsaritsyn and in Moscow, selling his own oil, prospering for the same reasons that Nobel prospered. He was interested in the problems of his people and became chairman of the city council.

Fellow Tatars respected and followed him, for no Mohammedan born or residing in Baku had ever been as successful. None ever did more for his people and city. Illiterate—it was Nobel's land agent who taught him to write his name—he established a newspaper, built the town's first theatre, organized schools and each year subsidized the education of a score of young Tatars by sending them to the local high school and then to various universities. He also started a school for girls, convincing the townspeople they should educate their daughters. In time he sold his oil company to the British, putting most of the 5 million gold-ruble profit into a textile mill east of White Town. "Why send the cotton from Baku to Moscow where the factories use our fuel to manufacture and then ship the clothing back to Baku?" He brought machines and instructors from England to teach the Tatars the trade and art of weaving. Tagiev made certain that his mills produced great quantities of less expensive materials and then made sure these found their way to the poorer inhabitants of his native city.[9]

Tagiev's career was a refreshing contrast to so many of the other oil barons who used their sudden wealth to devote themselves to the wildest forms of self-indulgence—leaving nothing to be desired, as a son of one of these magnates admitted, "in the way of barbaric luxury, debauchery, despotism and extravagance."[10] Vast sums were spent for palaces to house these newly rich. One owner wanted his home built entirely of gold but had to settle for mere gold plate when the architects and authorities persuaded him of the dangers involved. Another had a palace constructed in the form of a dragon—the entrance was through the gaping jaws. Another built a three-story mansion in the shape of a house of cards with large golden letters proudly proclaiming across the front, "Here live I, Isa-Bey of Gandji."

Such excesses eluded Ludwig and the Nobel officials residing in Baku. Their Villa Petrolea provided as much luxury as a Swede in southern Russia could hope to find. There were too many more important things on which to spend money: new lands, pipelines, derricks, refinery capacity, tankers, meeting rooms. The list was endless and the money continued to flow from the Nobel treasury. In 1886 the dividend had to be omitted and there was a revival of the rumors that previously had plagued Ludwig. Exaggerated by the competition, these reports of financial difficulties, overextension, and threatened bankruptcy soon reached Alfred in Paris. He promptly resumed his attacks on the company management, questioning once again the too-rapid expansion, doubting those optimistic predictions of future growth and prosperity.

Lagerwall served as his informant, complaining to Alfred that the company was still in its "Beliaminsky period," that overconfident mood of constant expansion with the overbuoyant Beliamin enthusiastically seconding all of Ludwig's grand plans. He wrote Alfred that overproduction in 1884–1885 had begun the problems, that the company had stretched itself too thin by paying high dividends in those years, that investment in ship and barge construction was much greater than budgeted, that Ludwig was considering a new loan for an additional 3 million rubles. There was also trouble in Tsaritsyn, Lagerwall reported; General Nyberg, recently placed in charge of that complex, had badly mismanaged affairs and had to be replaced by the Russian Gubischitsch. A few months later Karl Wilhelm Hagelin was given the post, the first important step up the Nobel ladder for the man who would play such an important role in the future fortunes of the company.[11]

Alfred might have found some comfort at this time by the receipt of 38,000 German marks, a royalty payment for dynamite sold in Russia from his Hamburg factory. Total sales in Russia since 1880 had been only 750,000 marks, an insignificant figure when compared to what Alfred was earning in other countries, especially from those French and British trusts he had recently organized. He probably blamed Ludwig for the sluggishness of Russian sales; certainly his mood was such that he could perceive little good coming out of Russia or from his brother's efforts.

Ludwig was the local Nobel dynamite agent; for several years he had been giving lectures and organizing demonstrations, arranging for tests, filling the ears of the military, encouraging the authorities to build a Russian dynamite factory, to become independent of foreign suppliers. He and Alfred were half-owners of the consortium created to construct that factory; Robert Liander, another Swedish expatriate resident in Petersburg, and Isidor Franzl, who had the concession for the Polish plant, were their partners.[12] But they never built the factory. The Ministry of War conducted extensive tests in 1880 but their favorable reports were submitted at the time of an outbreak of terrorist attacks. The assassination of the tsar in the following year ended all discussion of putting a dynamite factory inside the borders of the empire. Ludwig wrote Alfred that developments were at an impasse, that he was pessimistic about the chances for a factory; but Alfred was not easily pacified. At one point his disgust with the situation and his overwhelming pessimism about the future of the oil industry prompted him to consider accepting Ludwig's desperate offer to mortgage his Petersburg home as security for Alfred's loan to Nobel Brothers.

That home was worth a few million. Ludwig had just added another wing, extra space for offices—until 1910 all the oil company offices were located in the building. Ludwig had worked closely with the Swedish architect who drew up the plans for the major expansion and reconstruction of the house in 1874 and the result—an imposing three-story Florentine villa—was a most impressive addition to the recently completed Sampsonievskaya Nabereshnaya, the Sampson Quay, overlooking that arm of the river Neva called the Bolshaya Nevka. The ground floor contained the offices of Nobel Brothers. The floor above, the bel-étage, had large salons, a ballroom, music room, a dining room with a table seating twenty-eight, and several family rooms—along with Ludwig's office. In the top story were additional rooms for the family, several for guests, a billiard room, and the kitchen—the food was sent down by lift and there was a narrow staircase from the kitchen to the pantry below. A large winter garden two stories high connected Ludwig's office with the billiard room. The entrance to the offices and to the factory was behind the main house where there were also stables for a dozen horses, carriages, cars, and sleds. A mortgage certainly would have provided the cautious and worried Alfred with all the collateral he could use.

But talk of mortgages from a brother was not what Ludwig needed. He was beginning to feel the strain of the struggle—the jealous, unyielding competition, the still not fully measured threat from the Rothschilds, the obstinacy and stupidities of the bureaucracy, the reluctance and short-sightedness of the banks, and now the recalcitrance of a brother—a surfeit of challenge even for the man who thrives on it.

Ludwig thrived and, despite failing health, for a time survived. When the Volga thawed in the spring he and Lagerwall headed for Baku, inspecting installations, discussing future plans. In June he was again in Baku, this time with Jules Aron who had been sent by Rothschild to continue the dialogue and to organize a new refinery Bnito had recently purchased. In September William Herbert Libby, minister plenipotentiary from the House of Rockefeller, made the pilgrimage to Petersburg to discuss cooperation with Ludwig. A flattering gesture and recognition of the Swedish pioneer's power in the industry, but the contact produced nothing more than friendly exhanges. When rumors of war circulated in the capital Libby hastily withdrew. The Standard emissary had seen and heard enough. Ludwig was not currently in a position to control the total export of Russian oil; without that control, that guarantee, an agreement with Nobel alone for the purpose of market and price control would not interest Rockefeller. Libby also could report to New York that there was no immediate danger of Russian oil flooding world markets; the lack of adequate transport to Batum, the lack of a Baku-Batum pipeline, would continue to benefit Standard.[13]

Another visitor in 1886 was Fred Lane of the London firm of Lane and Macandrew, the shipbrokers who had imported the country's first Russian kerosene. It was a one-time affair with Nobel; in subsequent years Lane worked with Nobel's primary competitors. He worked with so many competitors, in fact, he was sometimes called "Shady Lane"—it often was difficult to discern just whom he was representing in a particular discussion or negotiation. But this blond and burly, personable, keenly intelligent individual was as honest as he was capable and he had a great capacity for work and considerable charisma. He was, in fact, the father of the British oil industry, an expert in all phases, especially transportation. It was Lane who arranged for tankers to carry Rothschild oil across the Mediterranean and it was Lane who organized Rothschild's London companies and it was Lane who inspired and guided the officials of Shell; finally, it was Lane who worked with Royal Dutch and helped organize the British Petroleum Company (BP).[14]

During his 1886 visit he shipped oil from Batum to India. A few cases of Russian kerosene had reached the subcontinent the previous year, but Lane's cargo was the first important shipment. Two years later the first Russian kerosene arrived in Singapore. In 1890 Lane returned to the Black Sea port, this time with a noted British merchant-trader Marcus Samuel, who was interested in purchasing Bnito's case oil for sale in the Far East, a market up to that time completely monopolized by Standard Oil. Son of an East End family that had started its struggle for survival by selling small boxes made of varied seashells, Marcus and his brother had accumulated a modest fortune from their trade with the Orient. Half of all Japanese rice exports were shipped with a Samuel stamp. It was the kind of bulk business Samuel understood.

In Batum there was another. The tankers of Nobel and Suart filling their holds with kerosene; the smaller vessels loading cases—all transporting their precious cargoes across miles of open sea. Why not larger, better-built tankers carrying Russian oil the fifteen hundred miles from Batum through the Suez Canal and the forty-five hundred miles to Calcutta, the five thousand to Singapore, seven thousand to Shanghai—to the ports Samuel knew so well where cheap local labor could be hired to build storage tanks, to manufacture tins and barrels, to organize transport and distribution into the interior?[15]

It was a bold, imaginative scheme, but one patterned completely on

Ludwig's transportation, storage, and distribution network. To beat Standard at its own game of price-cutting two conditions had to be met: a constant supply of bulk oil that was less expensive at the source than American oil, and availability of large and efficient tankers that could pass the safety regulations of the Suez Canal. In 1891 the first requirement was satisfied by the signing of a nine-year contract with the Paris House of Rothschild which agreed to supply Samuel with Russian oil from Bnito. Meeting the second condition was no more difficult. It had been the London Rothschilds back in 1875 who had raised the millions enabling the British to gain majority control of the canal and it was now the British House of Samuel using Rothschild oil and asking the British-controlled Canal Company to approve passage of its British-built tankers through Europe's new lifeline.

Standard had been the first to consider the idea—at least, they had been the first to apply for passage. But their tankers were too old, they were told, and not safe enough to prevent explosion and fire. Lane and Samuel made certain that the sleek new ships being designed by Sir Fortescue Flannery at Newcastle, West Hartlepool, and Sunderland would not be rejected; they were constructed in strict accordance with the new January 1892 safety regulations. There was thus no problem with the canal authorities when in May of that year Fortescue's *Murex* made its maiden voyage carrying four thousand tons of kerosene from Batum to the ports of the Orient. Standard Oil was as shocked as Nobel and the other Russian competitors of Bnito. Samuel, Lane, and the Rothschilds had kept their tanker plans secret and by the time the other companies recovered from the voyage of the *Murex* a sister ship, the *Conch*, was sliding down the ways. There soon followed a succession of others, all named for shells—Samuel never forgot those old shell boxes and the humble East End origin of the family.

Lane and Samuel planned well. Despite the opposition of those who continued to protest that it was unsafe to use the canal, despite Standard's vicious price-cutting efforts all across the Orient in one market after another and the cutthroat tactics of sales agents and near guerrilla warfare in their desperate efforts to retain their sales areas and the countless difficulties in organizing a Nobel-type storage and distribution system in sometimes primitive economic and commercial conditions—despite all this, Samuel prevailed. At one point the tsar himself brought pressure against the British for refusing to allow tankers of the Black Sea Steam

Navigation and Trading Company to use the canal—the tsar's family were the largest shareholders in that enterprise organized by Grand Duke Constantine. On another occasion Standard negotiators were discussing with the Rothschilds in Paris a scheme to strengthen Bnito in its struggles with Nobel in exchange for an agreement dumping Samuel and working with Standard in the Far East. But Samuel was not dumped and his Tank Syndicate organization prospered, eventually becoming the Shell Transport and Trading Company. Fifteen years after the *Murex* steamed through the Suez Canal more than two million tons of oil had been shipped the same route, 90 percent of it in ships of the Samuel line.

In the years before World War I Shell would play an increasingly important role in the Russian industry; when joined with Royal Dutch it would assume a prominence to rival Nobel. Royal Dutch–Shell then would be under the command of that Napoleon of the industry Henri Deterding, who brilliantly eclipsed Samuel and assumed control—with the ubiquitous Lane acting as middleman—of the joint venture. But by that time Samuel, his major successes well behind him, was occupied with the performance and perquisites of his newfound wealth and position. His sons were at Eton, he was elected to the Carlton Club, he had a town house on Portland Place, a country estate in Kent, was knighted by Queen Victoria, and became Lord Mayor of London. As capstone, he was presented the mantle of Lord Bearsted.

Both Samuel and Deterding represented a new generation of the oil industrialist, a new breed far removed from the pioneering do-it-yourself spirit of Ludwig Nobel. Deterding did not start working for Royal Dutch until 1896 but long before that, before Samuel ever saw those Nobel tankers in the waters of Batum, the third generation of Russian Nobels was sitting in the chair of command on Sampsonievsky Prospekt.

Ludwig, after another struggle with his health and several months of enforced rest in Petersburg and then on the Riviera, died at Cannes on 12 April 1888. His bronchial attacks had increased in severity; he was suffering from angina pectoris, and this time escape from the harshness of Petersburg's unrelenting winter did not bring the needed relief. His state of exhaustion was total, his health was shattered. His heart failed. He was in his fifty-seventh year.

There was a long line of mourners on the route to Petersburg's Smolensky Cemetery and many columns of praise in the press and on the lips of family, friends, business associates, and competitors, but none were more fitting than that offered a few years earlier when Charles Marvin first came across the man he crowned the "Oil King of Baku." Ludwig Nobel was a man of "total integrity," an "engineering genius" who combined his "inventive talent" with a great "capacity for organization" and the "power of patiently pressing down obstacles, and by sheer force of character commanding success."[16]

CHAPTER NINE

Divide the World

Some of the European press, confusing the news from Russia, reported that Alfred Nobel had died. From his laboratory at Sevran-Livry a few miles outside of Paris the inventor of dynamite had the distinct displeasure of reading obituaries condemning the munitions maker, the warlord who had made so much money finding new ways to maim and kill. Ironically the mistake was harbinger of history's treatment of the two Nobels: Ludwig was soon forgotten; Alfred, brooding about his own obituaries, eventually rewrote his will leaving all his fortune to purposes which no obituary would ever condemn and which would guarantee forever his fame.

Ludwig's estate was left to his wife Edla and to his ten surviving children. Emanuel, the eldest at twenty-nine, assumed family responsibilities in Nobel Brothers, and twenty-six-year-old Carl, the second oldest son, was put in charge of the factory—owned outright by the family which also controlled the vast majority of shares in the oil company. Of the 250-ruble shares 11, 211 were in their hands compared to Alfred's 4,838, Beliamin's 701, Robert's 240, and several other shareholders who had from forty to a hundred each.

Emanuel was just a year older than Ludwig had been when he took charge of Immanuel's bankrupt factory and he had about the same amount of on-the-job training. Ludwig, never a believer in classroom education at the expense of actual experience, had put Emanuel to work when he was sixteen, apprenticing him to a local factory where he worked a shift from six o'clock in the morning until midnight. Even Ludwig, with all his capacity for work, was impressed by those hours and by his son who managed to stick it out for a year. He had learned early the satisfaction derived from sustained periods of hard labor. "The main thing is the work," he declared. "The victory of enterprise and initiative is worth more than money."[1]

One of Emanuel's companions in that early factory training was another Swede, Hjalmar Podalirius Crusell, three years his senior and a gifted technician with similar belief in the gospel of labor. He spent many years in Baku installing the continuous-distillation refinery and then the first benzine-distillation plant in Russia. In 1885 he was put in charge of the company's technical section, holding that position for thirty years.

Emanuel's initial visit to Baku had been in the spring of 1876 when he accompanied his father on his first trip. Ludwig left him there for several months; it was just after he had completed the arduous apprenticeship and he needed the change of scene to rest and recuperate. Ludwig also told him to learn all he could about the area and the oil industry and later, back in Petersburg, he drew on his son's knowledge in organizing the new enterprises.

For Emanuel the months in Baku provided a solid introduction, and for the next half-century he was never far in thought or spirit from the shores of the Caspian Sea and the installations and activities of Nobel Brothers Petroleum Company. In 1877 after his return from the battlefield of Plevna, where he had delivered the infantry shields to General Todtleben and earned his St. George ribbon, Ludwig assigned him administrative and financial responsibilities for the expanding operations in Baku. When Nobel Brothers was established two years later he was given parallel tasks; after the reorganization in 1881 he became a director in charge of the department dealing with finances, government relations, and the stock market. It was early recognition of Emanuel's special talents and in the difficult years of the 1880s he proved his competence, earning high marks for securing the precedent-setting loan against future delivery of kerosene. At first only the private banks agreed to Emanuel's terms, but finally he convinced the State Bank to advance credit on the same basis.

Neither inventor nor engineer nor one to follow his father's intensely personal method of control, Emanuel with his special gifts resembled more his uncle Alfred. He too was destined to remain a bachelor but, unlike Alfred, Emanuel was usually surrounded by people; with so many younger brothers and sisters he inherited at his father's death both a large family and Ludwig's role as head of that family. Deep blue eyes, in later years a distinguished white beard, an outgoing and friendly manner, and a genuine regard for his fellow man—these were characteristics he shared with his father, not his uncle. But in 1888 Emanuel and the company needed the financial and business acumen of Alfred. The empire that

Ludwig had built was only a foundation. As the investments and innovations began to bear fruit greater management skills were required and a seemingly never-ending supply of capital. It was an industry sucking monies into those wells as fast as the black gold was streaming out.

The directors of Nobel Brothers could not agree as to the best means of raising those monies. With the central power suddenly removed there was open disagreement on future courses of action. Ludwig was gone and there was no one to take his place as a promoter of unity, as the one to make the final decisions. Emanuel, even as the family representative in control of the majority of shares, did not feel that he could make those decisions or that he could wield a strong hand over the board. This lack of decisiveness was a failing that troubled both him and the company in future years, but in the first months after his father's death it was no doubt a wise and cautious means of handling the older and more experienced members of the board. Emanuel deferred to those members, notably to the company president Mikhail Yakovlevich Beliamin, the sixty-year-old engineer who had been manager of the company which had taken over Immanuel's factory. Lifelong friend of Ludwig and—like him—an active member of the Petersburg Technical Society, Beliamin had been a supervising engineer in the Ludwig Nobel factory and in the early days of the oil company had proved extremely useful helping in the design of pipelines, tankcars, ships, and barges. When Robert departed he was appointed to take his place on the board. Whatever his shortcomings he was a native Russian, a qualified engineer, a reliable, tested friend and ally. That breed was not easy to find.

Beliamin succeeded Ludwig as president of Nobel Brothers and was also in charge of the sales department, replacing Gysser who moved to another firm. The old Russian was out of his element in sales and soon was confronted by Lagerwall with ambitious and generally progressive ideas for modernization and reorganization of the sales department. He proposed the division of sales into domestic and foreign sections but the board, led by Beliamin, rejected the idea. Lagerwall then presented a fourteen-page memorandum detailing a basic restructuring of the sales and distribution network in England, attacking the proviso in the company's British agent's contract that all oil—kerosene as well as lubricating products—had to be delivered in barrels. Why no mention of bulk? Why not a return to bulk deliveries in large tankers? Nobel had pioneered the idea and had clearly proved its value. Why abandon it?[2]

Lagerwall also accused Beliamin of financial mismanagement, blaming him for the needless loss of 200,000 rubles on the foreign exchange market. But the Russian was immune to these attacks and they only made him more autocratic. When Emanuel refused to support Lagerwall or to challenge the authority of the president an open split threatened. Lagerwall was furious, dismissing Beliamin as someone who "makes algebraic formulas out of the simplest matters" and the young Nobel of making a fetish of the old maxim, "To err is human and to forgive divine." He also charged that Emanuel had "a great fear of change," and suggested that "he should cast the *status quo* in bronze."[3]

What Lagerwall failed to understand in these first months after the death of Ludwig was the necessity for Emanuel to move with caution within his own staff. He was not exactly the inexperienced young lieutenant newly arrived in camp, but he was half the age of Beliamin and obviously not eager to alienate him and perhaps lose his Russian presence on the board—it was required by the local laws of incorporation—any more than he wanted to lose any other veteran. Lagerwall seriously feared for the future of the company and he was desperately impatient with Emanuel's indecision. He advised Alfred a year after Ludwig's death that Emanuel's good nature and "lovable personality" made him too easy a mark for someone like Beliamin to control, an open prey for someone else who could come along and really ruin him and the company. He pleaded with Alfred to write his nephew for "no one other than you, Mr. Nobel," could change the present situation. "Emanuel is suspicious about the advice of others but he will listen to you."[4]

Alfred had his own problems. He had granted the Italians the rights to manufacture his newly patented ballistite and the French, who had earlier refused his offer of those rights, ran him out of Paris. He was in the process of establishing a new laboratory at Mio Nido, a villa in San Remo. He was engaged also in a bitterly contested patent fight with two British officials to whom he had confided his ballistite formula—they promptly started producing the same product under the name cordite. But most important, Alfred was overwhelmed by a herculean effort to clear his name and salvage all his works from the aftermath of a great scandal, the work of a trusted French associate and the *cause célèbre* of the press. He had little time for the problems of his young nephew.

Emanuel really did not need the guidance of uncle or the presence of father. He was feeling his own way, using his own instincts and experience in trying to guide the destinies of the empire. And he eventually found his way. Four months after Lagerwall had expressed such great concern he was writing Alfred that it was no longer possible to regard Emanuel as a "negligible quantity." Emanuel had made some excellent official contacts, had done an outstanding job in organizing company and family finances, and had managed to curb Beliamin's power. To Lagerwall the future finally seemed secure and he recorded with some pride that Emanuel had done it all alone: "It has made him into a man."[5]

It also made him into a Russian. Tsar Alexander III personally made the request during an official visit to Baku. Ludwig had long anticipated and planned for such a visit but it was left to his son to do the honors. Fields and refineries were viewed and there was a demonstration at dockside—loading the tanker *Darwin* at a rate of a hundred tons an hour without spilling a drop. Emanuel presented the tsar's children with mementos of the visit, miniature silver drilling towers complete with pump house. The official party spent so much time at the Nobel installations that they had to cancel scheduled visits to other firms, much to the annoyance of Emanuel's competitors.

It was after the tour, during the exchange of champagne toasts, that the tsar said he hoped to have the satisfaction of seeing Emanuel become a subject of the emperor. The young man immediately replied that he had waited for just such a moment to ask for that great honor. Emanuel, the perfect host and already a polished diplomat, thus became the first and only Nobel to take that step. From the court's point of view it was a logical move. Why shouldn't the leader of one of the country's most important enterprises be a Russian citizen? For the Nobels it was also logical. Emanuel's education and experience were Russian; becoming a citizen might help dissipate some of the jingoistic prejudices swelling in the land and stimulated from the highest circles. Besides, one does not easily say no to a tsar. "God himself commands that his supreme power be obeyed out of conscience as well as fear"—so read the fundamental law which declared the emperor to be "an autocratic and unlimited monarch."

Two years later there was another official visitor. Finance Minister Vyshnegradsky arrived in Baku. Emanuel was again the host—he excelled in the role. He toured the minister through the Nobel complex, proudly pointing out the newest Nobel gusher which produced more than three hundred twenty thousand tons that year. As leading industry spokesman he used the occasion to plead for official sanction of his efforts to organize that industry, to combine the marketing and distribution networks of many of the competitors into one syndicate. Vyshnegradsky was receptive to the idea and he was impressed with Emanuel. The following year he appointed him to the State Bank's Discount Committee, an important post and a significant display of confidence by the government.

Emanuel thus was entitled to be addressed as "Your Excellency," and he held the post until 1918.

The two years after the minister's visit were known as "the hunger years." Severe famine spread through the grain provinces of the south and along the Volga. The price of Baku crude collapsed from 8 kopecks a pood to one. Kerosene fell 7 kopecks a pood, or about a fourth of what it cost the smaller refiners to produce it. Land prices dropped and several of the pioneers, including Kokorev, sold out. A half-dozen refiners offered their installations to Nobel but Emanuel was hesitant to commit additional funds. He had just completed construction of new storage tanks and was in the process of filling them with more than a half-million tons of crude and twice as much kerosene at the deflated prices. He was concerned about the increasing competition abroad from Standard, naturally taking full advantage of its recently established transatlantic bulk transport and distribution system.[6] Also, the Rothschilds were increasing their investment in Russia; they purchased the majority of shares in the Mazut Company, Nobel's largest domestic competitor. Mazut agents, inflated by the infusion of Rothschild interest and investment, suddenly were swarming all over the country in search of new markets.

By 1892 most of the purchase agreements Rothschild had negotiated when advancing funds for completion of the Transcaucasian railroad had expired. The producers in position to terminate those agreements did so and the Paris firm was forced to negotiate new contracts, to find new sources of kerosene to satisfy their own demand and that of Samuel and his Shell tankers sailing to the Far East. It was therefore the Rothschilds who did most of the buying of troubled refineries during the hardship years of 1891–1893. There was no shortage of funds for the purpose. In addition to the resources in Paris the local Bnito agents could use some of those profits realized by the previous sales agreements, those which enabled them to buy kerosene in Batum at Baku prices—the price differential often approached usurious levels.

There was plenty of oil available. The gushers kept pouring it out. For the smaller independents it was a potentially ruinous situation and in order to escape the clutches of Rothschild some type of cooperation obviously was mandatory. In 1892 Nobel and six of the leading refiners, excluding Rothschild, gathered in Rostov and organized a syndicate giving Nobel the exclusive right of distribution and sales for a period of five years. The combined productive capacity was eight hundred thousand tons and it was precisely the type of joint marketing arrangement and pooling of assets that Emanuel had proposed to the Minister of Finance.

Others also recognized that cooperation was long overdue. One of those was Calouste Gulbenkian, that shrewd Armenian whose later exploits in Iraq earned him the title "Mr. Five Per Cent." The twenty-year-old King's College graduate visited Baku and published an expert assessment of the industry in a series of articles in the 1889 Paris *Revue des Deux Mondes*. It was the only time Gulbenkian ever saw an actual oil field but those articles, published later in book form, established him as an authority. He knew his subject well; his father and uncle had been importers of Baku oil for years and the father had pioneered the introduction of Russian kerosene in Turkey.

The cooperation sought by Gulbenkian and Emanuel was realized in Rostov but only for a short time. Signatures were still fresh on the agreement when the lawyers and individual refiners started arguing. In a few months the Rostov syndicate was dissolved. All the producers who were not part of it were as delighted as the government, which was not quite ready to handle the kind of threat to monopolize the marketplace implicit in such an attempt to bring the market to order. But within a year after the collapse of the syndicate the government was forced to face that threat head-on.

A new association was formed, this time including Rothschild interests and the Armenian Mantashev who had also increased his holdings during the time of troubles. After Nobel and Rothschild, in fact, he had become the third largest refiner and exporter, and by the terms of the February 1894 agreement was given exclusive distribution rights in Egypt, Palestine, Syria, and India. Nobel was limited to Europe and Bnito was allowed to retain its own network and to honor its current contracts. A producer in Baku could thus choose which of the three to sell to—Nobel, Rothschild, or Mantashev—at a price regulated by a fourteen-member commission in Baku made up of both the major and minor refiners. It was a giant step forward for the industry, a serious, rational effort to move from the morass of jealousy and rivalry that had for so long prevented any form of cooperation. Not surprisingly, Standard studied the association with great interest.

In 1886 Standard had rejected consideration of an agreement with Ludwig because he could not bring in the other producers; Russian marketing conditions and the fierce competition were too chaotic for the Americans. But the new association seemed to indicate that some order now was emerging from that chaos, that some semblance of control might be possible. With Russian oil flowing from the ground in ever-increasing quantities and the rate of Russian production climbing faster than the

American, Standard was again haunted by the specter of Baku oil flooding world markets and costing only a fourth to a third as much to produce as the American. The able ambassadors of Rockefeller needed no encouragement to discuss possible cooperation. But just to make sure no opportunity was missed Alfred did his bit to bring the two parties together.

The brain which had organized the dynamite trusts had an abhorrent fear of an all-out struggle with stronger competitors and he single-mindedly decided to propose to Standard that they buy into Nobel Brothers, taking 49 percent of the common stock or 15 million rubles of a new issue. Standard quickly dispatched an emissary to Petersburg to discuss the proposal, but Emanuel was not willing to part with half his company. Having 16,289 of the company's 20,000 shares in his pocket, what he decided was obviously more important than what Uncle Alfred wanted. But Alfred was never one to surrender easily. When the Petersburg mission misfired he turned to Rothschild, encouraging them to consider alliance with Nobel. That seemed to be an unlikely prospect so he returned to Standard and held a series of meetings with their crack diplomat, the indefatigable Libby.[7] But not even Libby and an Alfred Nobel could meet the demands of Emanuel. Speaking with a newly discòvered voice of unity Nobel asked if Standard could control the export of the other American producers. It was a page from the Standard manual of operations and the company was as helpless in the face of its sudden American competition as Nobel had been in 1886. The Mellons were selling refined products to England and negotiating with French refiners; the Producers Oil Company, later reorganized as the Pure Oil Company, was selling to Germany. Standard's monopoly had been broken.

In June 1893 *Bradstreet's* reported that Standard was negotiating with Nobel and the Rothschilds, working out "a scheme for parcelling out between them the whole of the refined oil markets of the world"—the Russians would have the monopoly in Asia and Standard in England and Germany. The report was premature. Standard still had to make one more effort to control (usually defined in their manual as "eliminate" or "absorb") their American competitors. In the fall of 1894 they launched another price-cutting campaign. It was only partially successful but the effect in Russia was profound. Prices fell in Batum as the smaller packers and shippers were driven to the wall. When the dust settled Mantashev, who again had been the most ambitious in buying up the troubled independents, emerged stronger than ever and was for the first time in a position to challenge Nobel and Rothschild. But it was Nobel Brothers that reaped the first real profit from the newfound order and consolidation. After being forced to omit its dividend in 1893 (5 percent had been paid in both previous years), 6 percent was paid out in 1894 when the company showed a profit of nearly 3 million rubles. Rothschild that year lost 1.2 million.

In order to ensure future profit from the benefits of cooperation Nobel took the lead in reopening discussions with Standard (whose dividend rate at the time was 12 percent). The *New York Herald* headlined the meeting as a secret division of world markets. Again the press was premature; again Standard, despite all its price-cutting campaigns, was unable to drive its American competitors out of Europe. Plans to divide the world were indeed made and the agreements actually drawn up. But Standard had to admit defeat. It could neither conquer nor control.

Equally important was the attitude of the Russian government. After giving its official blessing to the conference the government reversed its policy and worked to defeat any agreement. Freight rates were lowered on the Transcaucasian railway; Batum oil became less expensive and even more of a competitive threat to Standard. The action was applauded by the European countries fearing a Standard monopoly. The *Berliner Tageblatt*, reflecting its government's reaction, wrote that the Germans felt far safer in the hands of Count Witte and the Russians than in those of Standard and the Americans.

Negotiations broke down soon after the 14 March 1895 preliminary agreement was drawn up between Emanuel Nobel and Jules Aron "on behalf of the petroleum industry of Russia" and William H. Libby "on behalf of the petroleum industry of the U. S." That treaty covered crude and refined products and assigned to Libby, Standard, and the American industry 75 percent of the world export trade and to Emanuel, Aron, and the Russians the remaining 25 percent. It was to remain in force ten years, but there was a stipulation that after one year it could be cancelled by either party with six months' notice.[8]

Nobel, Aron, and Libby—three men dividing the world. It was a scheme worthy of the age. Emperor and plenipotentiaries deciding the fate of nations and companies around the globe. And it was a scheme worthy of Russia's new finance minister, Count Sergius Witte. Dominating those around him by physical size as well as by personality and force of intellect, he was energetic and brusque, dictatorial and resourceful, and had a background which made him sympathetic to the demands and business affairs of the many Germans and Swedes in the empire. He was more

European than Russian and probably preferred the company of foreign capitalists to the reactionary corps of courtiers and Petersburg tsarlets.[9]

Emanuel enjoyed an excellent rapport with the tireless minister, supporting wholeheartedly his successful efforts to stabilize the Russian ruble, put the country on the gold standard, and build the Transsiberian railroad. All of these measures, of course, benefited Emanuel and the Nobel Companies; his tankcars could travel to the easternmost reaches of the empire burning Baku fuel oil enroute, foreign capital would be attracted to invest in a stable gold-standard economy, engineers, trained technicians, businessmen would follow the foreign capital. And that was the cornerstone of Witte's program: the infusion of foreign capital. With the strong support of Tsar Alexander he was able to promulgate such a policy in the face of great opposition within the State Council and those forces of orthodoxy and reaction working from the shadows of ignorance and inefficiency.

He also had to face the opposition of the secret police. His system of factory inspection to oversee his new labor laws collided head-on with the Okhrana's elaborate network of spies and informers inside those factories. When the incompetent Nicholas II succeeded his energetic father the police and other dark forces of Russian officialdom were encouraged to increase their opposition to Witte and his progressive reforms. The patriarchal autocracy which had benefitted Witte was replaced by an oligarchy of careerist bureaucrats jealously guarding competing government departments, one more inefficient than the next and each plagued by intrigues for power and position, consuming time and talent that should have been devoted to running the increasingly industrialized state emerging as a result of Witte's reforms.

During the count's first decade in office only Sweden could match the pace of Russian industrialization. Railroad mileage almost doubled; oil production increased 250 percent with coal and iron output rivaling that growth. Independent trading companies multiplied in Moscow, Petersburg, and Odessa, bartering the growing stocks of raw materials coming from the Caucasus and other formerly remote regions of the empire served by the new rail lines. No other nation was laying as much track as the Russians. In the Caucasus coal production from the Tqibuli mines and manganese from Chiatura—opened the same year Nobel Brothers was organized—rose tremendously; 38 percent of the world's manganese was coming from Chiatura. The pace of Caucasian life was also quickening. A

Georgian journalist wrote: "Everyone has come to feel that the era when it was possible to live an insouciant, idle existence at the expense of other peoples' efforts has vanished completely."[10] There was new life in the Caucasus and a new class of oil industrialist, wine and. grain merchant, mining engineer, tobacco manufacturer, and thousands and thousands of workers.

Branches of European companies made their appearance, with many more organized by European capital. The British, who had long been dominant in the Russian textile industry, were slow to enter the mining field; but when the Donets Basin and the Central Ukraine coal and iron fields opened it was they who led the westerners in developing those deposits. Welshman John Hughes became the Russian Iron King—they named the town after him, Yuzovka (later Stalino, now Donetsk). Hughes had no parallel symbolizing British interest and investment in the oil industry and it was a good many years before that interest was sparked, despite the example of the German banks, the Rothschilds, and the efforts of Charles Marvin whose series of pamphlets on the subject discussed what he termed "the coming deluge of Russian oil."

It was left to Alfred Suart to provide the example. The man who had first brought British tankers to Batum demonstrated to his countrymen that the risks were not so overwhelming. He had already proved one important point in the industry: when that other Britisher, Marcus Samuel, was sending his first ship through the Suez Canal Suart had sixteen tankers moving oil around the world. In later years as Samuel moved Russian oil east Suart concentrated on the west. To keep his ships filled and to remain independent of Nobel and Rothschild he purchased the property of Zatourov in 1896 and founded the European Petroleum Company. It was the first British company in Baku. The following year the Russian Petroleum and Liquid Fuel Company was chartered when a group of British investors bought the Tagiev installations at Bibi Eibat. Then came the Bibi Eibat Petroleum Company, Anglo-Russian Maximov, Anglo-Russian Petroleum, Baku Russian Petroleum, and Schibaieff Petroleum—one of the largest with a 1.5-million-pound capitalization. In a few years the total British investment approached 7 million pounds and the combined production exceeded that of Bnito. A new and vital force had been added to compete with Nobel, Rothschild, and Mantashev, already in command of well over half the domestic market and three-fourths of the foreign. To the government the British success was a welcome counterforce; any greater control by Nobel and Rothschild might have tempted the authorities to eliminate the threat of private monopoly by establishing its own.

As Witte's state became more industrialized, as more and more new companies were organized, the demand for oil increased and with it the importance of the Nobel Company. To fuel the machines of the new Russian industrial nation, the ship and train engines, the furnaces of metallurgical plants, the factories and workshops, Nobel and all the other producers were pumping a steadily increasing stream of fuel oil—the mazut of Baku which was replacing kerosene as the principal product. Eighteen ninety-three was the beginning of the spiral. That year Nobel's transport of fuel oil on the Volga nearly doubled. By the end of the decade the price reached more than twice its 1892 low of 7 kopecks a pood. With its vastly superior distribution system, its greater storage and carrying capacity, this mounting consumption of fuel oil was of greatest benefit to Nobel. In 1895 it paid a record 10 percent dividend, increasing this annually until it reached 18 percent in 1899 and 20 percent in 1900.

These were the years for which Ludwig had planned and struggled. The company he had organized and guided was now reaping the profits as Russia became the dominant petroleum producer in the world. By 1897 Russian fields—those twelve square miles around Baku--delivered just over 45 percent of the world's crude oil but—reflecting the country's greater use of mazut—only 23 percent of the world's kerosene. In comparison, the United States with 48 percent of the world's crude produced 64 percent of its kerosene. The following year Russia produced more crude than the United States and by the end of the century Russian wells were pouring out more than half of the total world production.

Through the 1880s Nobel wells surpassed 16 percent of the Russian total and by 1890 exeeeded 20 percent. At the turn of the century the company's hundred seventy-one active wells were pumping one and a half million tons, almost 18 percent of Baku's total and 9 percent of the world's. Nobel was also purchasing thousands of additional tons. With over a hundred storage depots and an extensive rail and water network to supply the increasing demand for fuel oil, it had to be purchased. The recently completed eight-hundred-thousand-ton storage reservoir near Balakhany brought local capacity close to two million tons. No one else could even approach that figure. Nobel dominated storage as it dominated transportation. Of all the oil shipped on the Caspian during the last decade of the century Nobel ships carried between a third and a half. On

Mother Volga nearly 60 percent of all the oil was moved in barges owned or chartered by Nobel.[11]

At the end of the decade Emanuel became president of Nobel Brothers. Beliamin, gradually eased out of active participation in company affairs, retired and received the traditional Fabergé silver vase. Emanuel, the dominant, decisive force in the company, now had the title as well as the power. He had guided the company through some of its most difficult and most prosperous years and entered the new century confident of continued prosperity.

CHAPTER TEN

New Directions

A major reason for Emanuel's confidence in the future of the company was the ability of the staff serving in top positions of leadership. That most trying task, the identification and recruitment of qualified personnel, was no easier for Emanuel in the 1890s than it had been for his father or his grandfather. With the economy expanding under Witte's careful prodding, more technicians and engineers were educated within Russia and emigrated from other countries; but as their numbers grew so too did the demand for their services. Companies fortunate enough to find qualified local talent did everything possible to keep that talent satisfied and working; the foreign firms did everything possible to attract their own nationals. Thus, from the time of Immanuel the Nobel banner brought together a loyal and relatively large band of Swedish workers, technicians, and executives; but no Nobel was more fortunate than Emanuel in his inheritance of that talent or his selection of new faces. While using his own financial and diplomatic skills to further company fortunes during the booming years around the shift of the centuries, he had his boyhood friend Hjalmar Crusell in charge of the technical section, Knut Littorin in charge of sales and distribution in the important Moscow district, and two more talented Scandinavians—Karl Wilhelm Hagelin, another Swede, and Hans Andreas Nikolai Olsen, a Norwegian—heading the other key departments of the company, the Baku complex and the sales section.

The first Hagelin migrated to Russia along the same route and for the same reason as the first Nobel and when he arrived in 1858 he soon found employment in Immanuel's plant. From that bankrupt firm he went further east and found a job on Volga river boats—he was the assistant engineer on the *Tsar* which took Ludwig and Emanuel to Baku for the first time in 1876. Three years later his son Karl arrived in Baku seeking work. The nineteen-year-old mechanic's first job was that of pipe fitter and his lodgings were as modest as his salary—he slept in the kitchen of the Nobel shop foreman. He was bright, diligent, hardworking, and ambitious and he saved practically every kopeck he earned and studied every spare moment. He was determined to be an engineer and in 1883, traveling to his country of citizenship for the first time, he enrolled as an extra student at Stockholm's Royal Institute of Technology. When he returned to Russia two years later—he had only enough money for two years of study—he was assigned to the Nobel laboratory in Petersburg.

Ludwig had established this research center a few years earlier; it was devoted primarily to finding varied and more efficient uses for Nobel Brothers petroleum and he was actively involved in the several experiments. When the young Hagelin arrived Ludwig's chief concern was with development of natural-draft fuel-oil burners, regulated in a manner to leave no residue and to produce no smoke. He was thinking about the Russian navy, hoping to convince the naval engineers and the Ministry of War to use oil burners throughout the fleet, not just on the Caspian. Hagelin worked with a boiler from an old locomotive and soon came to the conclusion that a natural-draft burner would not be able to satisfy the varying speed requirements of naval vessels. He installed pulverizers and after a time made a similar installation on an old torpedo boat the navy had loaned for the experiment. With four burners blazing the torpedo boat broke all speed records. But the navy was not convinced. They knew it was a good system, but as long as no other major power was converting to oil they did not want to take the chance. Hagelin and Nobel were twenty years too soon.

The negative reaction of the navy disappointed the young engineer. But the hours spent in the laboratory, the discussions with Ludwig which often lasted late into the night, were not wasted. In 1886 Hagelin was moved to the company's technical section, given a princely raise to 150 rubles a month, and two years later was sent to Tsaritsyn where he served as technical chief. His first official act in that position was not a pleasant one: laying the silver wreath from the workers of the Volga district on Ludwig's coffin. Two years later he was again transferred, again promoted—this time to the position of deputy manager of all installations and operations in Baku. In 1892 he became manager. The boy who had departed as pipe fitter returned nine years later at the age of thirty-two in charge of the largest company on the shores of the Caspian.[1]

The Norwegian Olsen had had no similar apprenticeship with the company although he could boast a closer relationship with the family—he married Emanuel's sister Mina. When he joined the firm in 1896 he replaced the aging Beliamin as chief of domestic and foreign sales. Emanuel had been advised by his uncle Alfred a few years earlier to find a young Norwegian to take into the business. Alfred had been as impressed by his contacts in Norway as he was unimpressed by so many of the Russian businessmen he had met, but even so it was a relatively radical recommendation for the man who had once reflected on the fact that of those associates closest to him most of the honest ones were Swedes. He did not doubt the loyalty or integrity of men like Beliamin; he simply had no respect for their business acumen. Nor, Alfred complained, did the Rothschild representatives. He was greatly relieved therefore when in the summer of 1896 Beliamin made his farewell tour of Europe, introducing his replacement to Nobel agents across the continent: Alexandre André in Paris, who had grown quite prosperous with his contract for Nobel lubricants in Europe; Paul Wederkind of the Società Italo-Americano del Petròlio, Standard Oil's Italian subsidiary which also purchased Nobel products; and the leaders of Berlin's Naftaport, a generally ineffective competitor to Standard's brisk and well-led organization in Hamburg. Olsen learned during this tour, in fact, that Nobel's marketing structure was no real match for Standard anywhere in Europe and as soon as he returned to Petersburg he initiated reforms, doing in part what Lagerwall had proposed seven years earlier.[2]

Olsen did find the time to attend the All-Russian Arts, Crafts and Industrial Exhibit in Nizhni-Novgorod, the traditional city of fairs and one of the transportation hubs of Russia where East confronted West. Emanuel spared no cost in the construction and furnishing of the Nobel Pavilion, a stucco model of the khan's palace in Baku complete with minaret and a glass cupola over a massive diorama. The German-Russian artist Schilder was commissioned to paint a mural representing company activities across the length and breadth of the land: drilling derricks, pumps, pipelines, storage tanks, refineries, factories, docks, barges, rail lines, ships. All was included and Schilder needed the help of company engineers to plan and lay out the five-meter-wide forty-meter-long painting. Emanuel felt his 200,000 rubles well spent and was glad to add to that expense by commissioning Fabergé to strike a special medal to be presented to friends as a mark of the occasion.

He was fully aware of the fact that Count Witte wanted to impress tsar, nation, and the world with the progress of Russia's trade and industry but the grand opening—with Emanuel ready to greet Nicholas II, Tsarina

Alexandra Fedorovna, and Witte—was a near catastrophe. The official party was scheduled to arrive a half-hour before noon but Emanuel and his staff, in morning coats and standing in front of the palace pavilion in a sea of flowers under flags and bunting, were nearly swept away by a violent storm which struck at eleven. Cyclone winds brought hail, rain, smashed windows, tore apart pavilions, destroyed flowerbeds, ripped flags from their masts, sent thousands scurrying for cover. It lasted just a few minutes but the damage was enormous. Nobel's glass cupola blew down in jagged bits tearing holes in the great curtains over the tribune; flags and flowers were ruined, Schilder's painting damaged. The visit had to be postponed twenty-four hours and there were many who saw in the flash storm an omen as foreboding for the new tsar as that which occurred a few months earlier at his coronation when so many were trampled to death during the traditional distribution of gifts. The loss at Nizhni was not due to human stupidity, however, and was easier to repair. After frantic round-the-clock efforts Emanuel and his staff again were ready to welcome the tsar and his Minister of Finance.

The Nobel Pavilion and the tsar's visit were high-water marks for Emanuel and Tovarishchestvo Nobel. Six months later it was a happy and consoling memory as he battled against the sudden market drop in Nobel shares amid rumors of impending disaster. He was pushed to his limits defending the worth and stability of the company and the good name of his family.

The crisis was caused by the death of Alfred in December 1896. There was immediate speculation that the financial bulwark of Nobel Brothers had been removed, that plans for expansion, for improvements, had to be curtailed, that the company was in serious danger of going under. When the terms of that remarkable will were made public, establishing a foundation and Swedish-Norwegian committees to award prizes for extraordinary human achievement, the impact was even greater. Alfred's holdings in the company had to be liquidated in order to finance implementation of the will and to establish the foundation. The stock markets of Europe were rife with rumors of ruinous forced sale of Alfred's shares. Few if any investors or speculators really knew the extent of Alfred's holdings in the company but most assumed that millions were involved, probably enough to control the company.

Many of Emanuel's friends and advisors, fearing imminent collapse from a market panic, urged him to contest the will. But as spokesman for the Russian branch of the family he adamantly refused. He had been one of the executors of an earlier 1893 testament, was the eldest of all the nieces and nephews, and certainly the one closest to Alfred. He was at San Remo the day after his uncle's death making arrangements for a service, for the opening and reading of the will, and for transport of the remains back to Sweden. He had no intention of contesting the provisions or disputing his uncle's final wishes. On the contrary he advised the executor Ragnar Sohlman to be guided by the Russian words "dushe prikashchik"—to allow the will and his own actions to be "spokesmen for the soul." Emanuel personally followed his own advice, becoming the strongest sup-porter of Sohlman in his arduous task of execution. Sohlman in later years never was able to "recall without emotion the high-mindedness and courtesy he displayed towards me, though I had been appointed to repre-sent interests which might be regarded as running counter to those he represented himself."[3]

Sohlman was a remarkable choice for the position, a man of high pur-pose and singular character, a final instance of Alfred's genius and judg-ment. A graduate of Stockholm's Institute of Technology (during one term break he traveled to Baku aboard a new Swedish tanker moving along the canal-Volga-Caspian route to Nobel docks), he was employed for a time as chemist in Alfred's American company Hercules Powder; it was while working in the Swedish exhibit at Chicago's Columbia Exposition that he received word of Alfred's offer of a position as his personal assistant. From that moment—organizing Alfred's papers in Paris, supervising the lab-oratory at San Remo, setting up a workshop at the Bofors factory, and on through those trying times of establishing the validity of the will and then chartering the Nobel Foundation, formalizing committee credentials, launching the annual institution of the prize-giving ceremonies—from that time in 1893 when he joined Alfred he was intimately involved with the preservation of the Nobel name and the furtherance of the lofty goals and purposes envisoned by Alfred and then guaranteed by his nephew.

It was Emanuel who finally prevailed over the other members of the family, in particular the Swedish branch, and convinced them to accept the will as written. Robert had died four months before Alfred and his heirs—his sons Hjalmar, Ludwig, and son-in-law Count Ridderstolpe —did everything in their power to void it, to keep Alfred's money within the family. They filed court suits; they tried to freeze assets in France, Ger-many, and Great Britain; they enlisted the support of highly placed Swedish officials and played on the fears of those who maintained that Swedish academies were ill-suited to select world leaders in science and art; they found receptive audiences for the concern and resentment felt when Alfred named the Norwegian parliament, still under the Swedish crown, to award a peace prize. To some extent they also managed to enlist the support of a few members of the Russian branch of the family.

This disaffected trio mirrored in part the bitterness and resentment that had troubled Robert ever since his retirement and departure from Baku. He never stopped complaining; in 1885 he proclaimed that "everything that has been done since I left Baku is expensive and poor."[4] Living the life of a landed gentleman, suffering from deteriorating health and increasing eccentricity, he·could not help envying the gigantic financial and industrial empires his younger brothers had created, in particular the enterprise built on foundations he had originally laid. Then, too, Robert's sons and daughter did not have the financial security of the other Nobels and it could not have been easy to stand by and let all that money go to strange people in distant lands because they had written books or devised some new formulae.

They made every effort to convert Emanuel to their position but he would not budge. In March 1897 when they learned he was in Berlin—he was negotiating a 21-million-mark loan from the Discount Bank in order to have funds available to purchase Alfred's stock and cover any emergencies that might arise as the result of a forced sale—they rushed to meet with him but he refused to see them. He had made his position clear: he would not fight the will. He was interested in preservation of the family name and honor, in strict adherence to the wishes of Alfred, avoidance of open court squabbles, and retention of the family business.

There was a close interrelationship among these interests and when Sohlman visited Petersburg a settlement was finally in sight. He arranged for transfer to Emanuel—as representative of Ludwig's heirs—of Alfred's holdings in Nobel Brothers. The price paid was a bargain 2 million rubles; the stock was assessed at par value which of course meant a considerable saving for Emanuel. For the Swedish branch of the family Sohlman devised an equally advantageous settlement. He had known Robert's son Ludwig since Stockholm school days and had first met Hjalmar when visiting Baku. He thus was dealing with known quantities, and the compromise he worked out was payment to Robert's heirs of eighteen months' interest on Alfred's estate.

It was an equitable and fair arrangement, one that made it possible for Emanuel's advice to be heeded: let the last will and testament speak for the soul. But it took all of Emanuel's persuasive powers and a full measure of diplomacy and tact to convince Hjalmar, Ludwig, and their brother-in-law. In a dramatic February 1898 gathering of the clan Emanuel made a final plea to honor Alfred's wishes, to protect the family name. It was a lonely role to play. Family, friends, the press, even the king opposed him. Summoned to the palace, he was told by Oscar II that "it is your duty to your sisters and brothers, who are your wards, to see to it that their interests are not neglected in favour of some fantastic ideas of your uncle." Emanuel was not convinced. He replied, "I would not care to expose my sisters and brothers to the risk of being reproached, in the future, by distinguished scientists for having appropriated funds which properly belong to them." When the king persisted in his argument Emanuel could only repeat what he so firmly believed: "I cannot take it upon my conscience to try to lay hands on the money my uncle intended for persons worthier than myself."[5] The agreement was made, the will validated; the Nobel Foundation with its $8.6 million was established.

Emanuel not only had preserved the honor of the family name. By his actions he had guaranteed for posterity the memory and annual remembrance of the achievements of at least one member of the family. But in 1898 it took courage to oppose popular sentiment, to fight his own family, to brook arguments with the king. His Russian valet feared for his life; when he returned to the Grand Hotel from his audience at the palace and explained to his valet that he had strongly disagreed with the sovereign, the valet started packing their trunks in a panic that the police would soon be at the door. But Sweden was not Russia and there was room for disagreement with royal authority. Emanuel could take his time about leaving the Swedish capital, content with his role as "dushe prikashchik."

Arriving in Petersburg he found his new sales chief making impressive progress with reorganization of the sales structure. Olsen was convinced that all sales agents, domestic as well as foreign, had to be brought into a closer relationship with the home office, into companies owned jointly with Nobel Brothers. He started with the Moscow district where an independent firm had been selling the Nobel line for many years without startling success. Their contract was terminated in 1898, a new organization established under the control of the main office in Petersburg with a capable executive, Knut Littorin, in charge. In the first two years profits increased tremendously, far beyond the most optimistic predictions. What had been a weak link in the Nobel cross-country chain was strengthened to the point that it became the model for all other domestic districts.

In addition to the advantages of centralized control and direction the

Moscow office benefited from the new spirit of cooperation with the competition encouraged and formalized by Olsen. For years the Nobel agents in Moscow had been troubled with what Olsen termed "intense, bitter and senseless competition" with the Pollack brothers Mikhail and Saweli, local representatives for Rothschild's Mazut Corporation. That company had been organized to handle distribution and sale of the crude that Bnito's first manager in Russia was producing from his wells in Balakhany. Grigory Pollack, father-in-law of the engineer who ran Bnito's refinery, founded the company with a 4-million-ruble capitalization and with Rothschild as the largest shareholder. By the end of the century, when Mazut had established itself as a serious but distant competitor to Nobel, the Rothschilds had 80 percent of the shares; the Pollacks and the Deutsch brothers split the remaining 20 percent.

Agreement on the basic principles eliminated the price-cutting and other standard tricks of contemporary commercial rivalry and in place of the "intense, bitter and senseless competition" there were market-sharing arrangements, fixed prices, and increased profits for all. The example was driven home to all other Nobel representatives: Russia was the company's most important market; it was essential to the future vitality and prosperity of Nobel Brothers that market conditions remain stable. Mazut's agents had similar interests and basically so did the government, although Witte was not eager to foster an over-cooperation which might lead to merger and then monopoly. He kept close watch on the local market and on the performances of Olsen and Emanuel and their representatives, just as he did when they performed on the larger international stage.

Coordination with Mazut's Moscow representatives certainly could not be made without the approval of their masters in Paris, and long before Olsen ever suggested cooperation on the local level he made sure he had the necessary sanctions from Rue Laffitte. When Olsen proposed a conference on the subject Aron came to Petersburg. That was in June 1897. The four-year-old syndicate of Russian producers was dying a slow but sure death and the houses of Nobel and Rothschild had to discuss future plans for regulation of the export market. Aron expressed complete agreement with Olsen's philosophy that the overriding consideration in domestic and foreign operations had to be the element of cooperation rather than competition and it was apparent that the Russian oil industry was on the brink of a new era.

In February 1900 Olsen was in Paris for two weeks of discussion with Aron and his chief assistant Maurice Baer who had been put in charge of the Russian petroleum section when Aron was made chief of the overall commercial and industrial section. Three months later Aron was in Petersburg and went from there to Berlin with Olsen. The Norwegian was attending the annual meeting of Naftaport and inspecting the new depot. He had been present at the previous year's meeting, initiating an increase in capitalization from 1.5 million marks to 5 million in order to build the new storage facilities. The chief of Naftaport's administrative council was one Vasily Timiriasev, the official financial representative of the Russian government assigned to Berlin—Witte was keeping a watch on oil affairs in Germany and Emanuel was keeping his governmental relations intact. From there Olsen went to Paris for discussions with Baron Edmond Rothschild who invited him to dine and attend the opera while approving arrangements for a London meeting to discuss full-scale cooperation in the German and English markets.

Olsen's goal was complete reorganization of Nobel's sales offices and an elimination of that "intense, bitter and senseless competition" abroad that was proving costly to all and beneficial to none. To achieve this he worked toward a revival of that spirit of cooperation which had almost brought about a formal division of the world markets back in 1894. In the German market he planned to bring together Naftaport and Standard's German-American Petroleum Company; in the British Isles he wanted a total revamping with new arrangements similar to those he had previously worked out in Holland and Belgium where Standard's American Petroleum Company had invested in Nobel's Henri Rieth firm and was marketing Nobel kerosene. Olsen already had the approval of Baron Edmond; it was not difficult to arouse the interest of Standard. In keeping with his principles of cooperation instead of competition, of discussion rather than dissension and disruption, from his initial meeting with Standard representatives in Europe he had emphasized establishment of the best possible personal and professional relations. When he reached London in June 1900 he therefore knew he would be dealing with a known and friendly colleague. James MacDonald, director of Standard's Anglo-American Company and replacement as Rockefeller's European emissary for an ailing Libby, thought highly of Olsen and was as eager as the Rothschild interests to bring some order to the European markets.

Olsen represented Nobel, MacDonald spoke for Standard, while Aron and Baer were the Rothschild delegates. Aron also brought a lawyer, Tambour, and Emile Deutsch. A few officials of the German-American Petroleum Company were present along with director Zahn of Naftaport.

And in the wings stood Fred Lane, who by this time was clearly identified as Rothschild's man in London and key advisor on all matters pertaining to petroleum. Olsen had a healthy respect for Lane and described him as a "clear-thinking . . . unusually clever" individual with great "practical sense" and "an earnestness of purpose."

Lane supported Olsen's aim of unifying Nobel and Rothschild sales in England but he resisted bringing Nobel's agent, Bessler and Waechter, into the deal. There had been some difficulty between Lane and the British firm and the bitterness remained, but Olsen prevailed. For Bessler and Waechter there was no other choice. Their contract had expired and Olsen definitely was not going to make any settlement in London that did not include Nobel ownership. After ten days of discussion, with the meetings lasting from early morning until late at night, agreement was reached. Consolidated Petroleum Company was chartered. The parent firm of the British Petroleum Company was born with Nobel, Standard, and the Rothschilds all assisting.

The agreement covering Germany was neither as dramatic nor as formal but the results were just as effective. With Naftaport's newly expanded storage facilities, a booming Russian production which accounted for more than half the world's oil, and an ably aggressive leadership in Nobel's sales organization, MacDonald and Standard were certainly willing to listen to Olsen's overtures of friendship and cooperation and to give their approval to whatever unofficial agreements were needed to continue that friendship and seal the cooperation.

Olsen and Nobel were dealing from strength—or at least they thought so—fortified by the apparent support of Witte and the government which was listening to the demands of independent producers for establishment of an all-Russian export union to organize shipment (with official government rebates on rail freight and Suez Canal tariffs), distribution, and sales in the Far East. As the major Russian company Nobel would be the logical leader of the new organization, but until completion of the Transcaucasian pipeline there was no realistic hope of getting enough oil to Batum to launch such a project; time passed and Witte reconsidered—he did not really want to risk that ambitious a venture in the British sphere of influence. Through his insistence that the proposed union include *all* producers he effectively squashed the scheme, although meetings discussing the project continued to be held in Baku and Petersburg.[6]

Deterding of Royal Dutch was spinning his own web at the same time. He realistically had assessed the unlikelihood of Russian unanimity and the country's inability to organize such a grandiose scheme and he knew that any union led by Nobel would drive the jealous Rothschild interests into the arms of Shell. And that company, with its large fleet and its existent marketing network throughout the Far East, was vital to any plans for a Russian export union. Counselor Goulishambarov reported to Witte that Shell was "a very great power, which the Russian exporters will have to reckon with seriously."[7] The judgment was a cautious understatement. Several independents declared that if the Shell fleet could not be made available to the union they would withdraw their support. Withdrawal would result in a shortage of oil in Batum, and without Shell tankers whatever supplies were available could not be shipped.

Deterding and the indomitable Lane had the answer. Shell should sign with Royal Dutch before Witte's union could be officially launched; the Russian independents would then flock to the new alliance. With Russian oil added to that already available to Royal Dutch from its eastern sources a vast new distribution and marketing organization could be established. Standard Oil finally could be confronted and defeated. They managed to convince Marcus Samuel—Lane and Deterding together must have been difficult to refuse—and in June 1902 the agreement was signed to organize what became known as the British Dutch, the Asiatic Petroleum Company. Shell, Royal Dutch, and Rothschild each put up $1.5 million and had equal shares in the new company.

Samuel resisted the entry of Nobel but Mantashev and Goukassov later joined with half a million. Nobel's Star and Crescent brand of kerosene had a good market in Shanghai; elsewhere in the Orient Bnito's Anchor brand was more popular and Samuel, speaking from years of experience in the area, advised his partners that it would be difficult to allocate markets. Samuel also was worried about Nobel's contract with an independent to take nearly nine hundred thousand cases of kerosene, half of which was destined for the Far East. With the market already depressed Samuel did not want to see all those cases heading east and cutting into his own profit, much of which he earned from the shipping and storing of bulk and not case oil.[8] Mantashev and Goukassov were similarly saddled with advance delivery contracts and after a time they dropped out of Deterding's new association. It really was no loss to the Dutchman. So ably did he lead his new endeavor while working independently to promote the welfare of his own Royal Dutch company that he was not at all concerned about the defection of the Russian producers. He had Rothschild—a sufficient, guaranteed supply of Russian oil. That was enough.

The Asiatic Petroleum Company was a great success and devoured one of its parents like a Gorgon. In the next few years Deterding was able to outmaneuver Samuel and establish the clear and undisputed supremacy of Royal Dutch. The Englishman had several chances to avoid what Deterding was planning as an inevitable outcome but he was too concerned with other things—with his new wealth and status in British society. When Olsen and Aron went to London in November 1902 to discuss Olsen's plan to ship Galician oil to Germany the head of Shell was too busy to see them. For six days they waited but Lord Mayor Samuel never found the time to talk about a marketing arrangement which might have given him a strong enough weapon to fend off Deterding's later onslaughts.

Like Samuel, Nobel was being outflanked and had to be content with other victories on the international front. An important one concerned lubricants. Ever since the first Russian oil exports to Europe lubricating oils had been treated as distinct from kerosene and other petroleum products and the various representatives selling those lubricants had no desire to change their traditional ways of doing business. But Olsen had other ideas and Emanuel sponsored an August 1900 Petersburg conference to discuss new arrangements. It took a week to convince long-serving agents like Alexandre André of Paris, Max Albrecht of Hamburg, Charles Good of Anvers, that there could be greater profit in one large joint sales company. Although the Rothschilds were not as concerned with lubricant sales they had four representatives at the meeting including Emile Deutsch's brother Henri. Emanuel chaired the proceedings and was host at the banquet held at the restaurant Ernest to celebrate formation of S.A. d'Armement d'Industrie et de Commerce, known as SAIC. Headquarters were set up in Antwerp, capitalization was 15 million French francs, and the organization was split into two groups, one for the producers (with Nobel holding the greatest number of shares—together with Rothschild the majority) and one for the distributors who remained independent dealers utilizing their own storage and transport facilities.

Six months later Olsen was in London negotiating with the English-owned Schibaieff Company, a specialist in lubricants and for years a well-established label on the continent. It represented no real competitive threat to Nobel in Russia but the newly established board of SAIC complained that it was a threat in Europe. Olsen and SAIC representatives therefore met with Schibaieff director Wilhelm von Ofenheim, a shrewd Austrian who later played an important role in the Galician oil industry.

Agreement was reached rather quickly; Schibaieff was given a 14 percent interest and the new SAIC was further strengthened. Marketing in excess of a quarter-million tons a year the new organization was an immediate success and it continued to grow, bringing a sizable profit to Nobel in the years before World War I. After that war it was one of the very few assets left to the company.

Olsen stopped in Berlin on the way back to Petersburg. There he met with the local Standard representatives. There were no substantive discussions but Olsen never passed up an opportunity to sit down and talk with the competition, to make sure there were no sudden or serious obstacles to continued friendly relations. And he never passed up an opportunity to meet with his two closest contacts in the other camp, Aron and MacDonald. By January 1903 he felt that his rapport with the American was warm enough to speak openly about the possibility of even closer cooperation, about a scheme he had been considering to bring the two industrial giants together in a new corporation to be formed by an exchange of shares. MacDonald was very interested and like Olsen agreed to check with the directors before continuing the dialogue. The response from New York was positive and—much to Olsen's surprise—so was Emanuel's; he was not noted for his snap decisions on matters of such import.

On a wintry February day in 1903 Emanuel and Olsen met in Berlin at the Hotel Bristol with MacDonald and Libby—the latter now sufficiently recovered to attend such a crucial conference. The Standard spokesmen had numerous questions about Nobel's firm, the properties, the capitalization, credit, and balance sheet. They already had a great deal of information gathered regularly through the years and deposited in the various Standard offices, and as the discussion continued it became apparent that the matter of Standard purchase would depend on the appraised value of the Nobel shares. The current market price was high but still only twice par value, and Emanuel explained that they had been pegged at a low book value without taking the sizable hidden reserves into consideration. The shares were obviously worth more than the market price but just how much more no one really knew. Emanuel agreed to undertake a new audit of all assets and to meet again in Berlin as soon as the results were available.

Hjalmar Crusell was given responsibility for the audit—he was the only other company official brought in on the project. The job was completed in July and the result was a shock to both Emanuel and Olsen. The actual value of the shares was eight to ten times current market price. Even allowing for some over-enthusiasm when making estimates and then making the necessary downward adjustments there was no way to make that price per share attractive to Standard. Emanuel was certain of that. MacDonald had the anticipated reaction when he learned the news.

Standard made no counterproposals and neither did Emanuel. With his friend and supporter Witte dismissed by a government that was growing increasingly reactionary and irresponsible, approval for any kind of merger with Standard would be difficult. Approval would have been difficult enough with Witte in office. But the discussions had not been in vain. What had commenced in 1886 when Standard first called on Ludwig and was resumed in 1894 with the ambitious but abortive scheme to divide the world, again was renewed as Nobel and Standard were drawn closer together in their common efforts to bring order to a chaotic and all too often cutthroat marketplace.

CHAPTER ELEVEN

Karl Vasilievich Hagelin

Emanuel could concentrate on the summit meetings and work closely with Olsen—supporting his efforts to revitalize European sales organizations —because he was spared the time and trouble of keeping a close watch on developments in Baku, of doing what his father had been forced to do when creating the empire. He had taken seriously that advice of his uncle Alfred that "if you try to do everything yourself in a very large concern the result will be that nothing will be done properly."[1] And his gift for assessment and recruitment of key personnel meant that he could delegate tasks with full confidence that they would be competently carried out. Nowhere was this more important than in Baku where the manager had the most taxing and challenging job in the company—it was recognized as the central nervous system of Nobel Brothers. And at no time did that system have a more capable chief at its head than Karl Wilhelm Hagelin.

Karl Vasilievich, as he was known to the Russians, had had years of firsthand experience in Baku as well as in the regions of the Volga. He knew the docks, repair yards, railroad terminals, storage depots, the machine shops in Astrakhan, Tsaritsyn, Nizhni-Novgorod. As apprentice, administrator, manager, and then director he was a frequent visitor to those installations, refreshing that knowledge, renewing friendships. The value of his experience demonstrated itself when he was placed in charge of the Volga district. There was then a general feeling among company directors that they should not put more time and money into the water routes, that Nobel already had enough tankers and barges, that additional vessels could be chartered whenever needed. But when charter rates increased and the full impact of the government's tight control and manipulation of railroad freight rates was felt the Petersburg directors gave serious thought to expanding their fleet. They turned to Hagelin on the Volga.

Karl Vasilievich was a child of the Volga. His boyhood had been spent on the shores of Russia's great trade artery watching the endless stream of boats carrying corn from Podolia, Astrakhan sturgeon, tea from China, Siberian iron, Bokhara cotton, Persian carpets, grindstones from the Urals, samovars from Tula. He knew the towns and villages, the shippers, captains, engineers, the seasonal changes, depths of the channels, costs of chartering, the way to deal with ship crews. Under his command, with his certain knowledge of the need for independent barge transport, the number of boats flying the Nobel pennant gradually increased. The board was not completely convinced, however; there were many other demands for capital and Emanuel never wanted to overextend. Hagelin's new barge construction and purchases were carefully monitored. In only one area was he given full rein.

He had conceived the idea of converting one of the Swedish-built barges to a propeller-driven steam tanker. A propeller craft would be lighter, less expensive, and provide more storage space than the paddle wheelers then in use. But no one knew whether or not the propeller machinery—thus far restricted to sloops—could be adapted to a larger shallow-draft ship. Hagelin ordered the propellers and engines from Motala and put Frans Ekman, chief of the company's machine shop in Tsaritsyn, in charge of the project; another Swede, Fredrick Arvidsson, ran the machine shop in Astrakhan. An oversize hardworking native of Björneborg (Pori), Ekman started his Russian career as a machinist in Ludwig's factory. He was promoted to the position of section chief in the Izhevsk plant and then Ludwig sent him to the eastern shore of the Caspian, to Usun-Ada where the company had a government contract to build a distillation plant to provide for the workers then laying a rail line connecting the Caspian with Bokhara, Samarkand, and Fergana. After the desolation of Usun-Ada, Tsaritsyn was a welcome assignment. Ekman arrived about the time the new storage depot and barrel factory were being completed. Placed in charge of a small machine shop, in a few years he was supervising a major operation where barges and steamers were overhauled during the long winter layovers when the Volga was frozen and where new ships and barges were built for the Nobel fleet. Converting a Swedish barge to a propeller-driven tank steamer was no engineering problem for Ekman and his crew.

But there were a few anxious moments when the newly christened *Anna*, named for Hagelin's daughter, stuck fast in the shallow slip. Would it sink? Was it seaworthy? Was the strain too great for the propellers? Hagelin called for a tug to fix a rope and yank the new boat free. Three hours later *Anna* was en route to Astrakhan under a full head of steam, propellers whirring and all hands confident. Another twenty-four hours and Nobel's newest ship was loading its first oil, ready for the three-hundred-mile return trip to Tsaritsyn. There were no trial runs for the Volga's first propeller-driven tanker as Hagelin proved once again the truth of the old proverb that "the Volga is a good horse; it will carry anything you put on it."

His new vessel was faster than the old paddle wheelers, had a 25-percent greater carrying capacity, and cost less to convert than originally budgeted—always a sure way to please management. Petersburg ordered the conversion of four other barges. Hagelin and Ekman had done their work well, had made their mark with the board. The credentials of Karl Vasilievich as company expert in all matters of river transport were now clearly established and accepted. Within a year he was studying charts and maps and working with Russian engineers familiar with the area, organizing barge transport for the rivers of western Siberia—the Ob, Tobol, and Irtyshsk.

While he was preparing for an on-site survey he received word from Petersburg that he had been promoted to the position of deputy chief of Baku operations, second in command to Leonid Richter, former Russian artillery colonel—an individual of energy and expansiveness but one who interpreted his responsibilities as primarily those of social cicerone. "I drank more champagne during the three months I worked with Richter," Hagelin later recalled, "than during the rest of my life."[2] When the board realized that Richter was not the man for the job they sent out their general secretary, the tactful and diligent lawyer Rafael Nikolayev. But he soon was swamped with local implementation and monitoring of the many sales and export agreements that Emanuel and Olsen were negotiating. Hagelin, whose assigned area of responsibility supposedly was limited to technical affairs, soon found himself heavily involved in all areas of management and administration. Deciding to let Nikolayev devote himself exclusively to the new trade agreements, the board asked Hagelin if he felt he could handle the job of chief. Hagelin replied, "Am I not already the chief there?"[3]

White Town and Black Town had changed greatly since the young Hagelin first arrived in Baku. There were now more than a hundred thousand people. The Tatar section did not look much different but it was certainly more crowded and there was more smoke and soot and a stronger smell of oil. In the fields beyond, the forests of derricks were thicker and the refineries much larger. The oil barons had built palaces; the boulevards were wider, the docks enlarged, and there probably was more electricity in use than in any other city of comparable size in the world. The streets were as cosmopolitan as ever with Caucasian tribesmen in long pleated coats and jeweled sword belts, Jews with fur-lined caps, Persians with waistcoats and brown felt caps, Armenians and Russians, Swedes and Germans—now and then a visiting American or Englishman. Poverty and squalor still were the lot of most of the workers and travel to the interior still was a risky business without guards, even on the trains. An English engineer observed that rail travel was most unpleasant, with "the nauseating smell of unwashed humanity."

Heat, humidity, flies, and dust made the summers unbearable. Those who could afford it still fled to the mountains, to the hill station of Adjikent where the governor general and most owners and managers had summer houses cooled by mountain air and surrounding greenery. Richard Sorge, Soviet master spy of World War II, was born there—his German father was an engineer in Baku. Hagelin's eldest son also was born there but the new father saw little of him those first weeks. The morning of the birth Baku's director received a telegram: "Cholera in Baku." He returned at once.

Panic gripped the town as the news spread. The railroad station sold forty thousand tickets; thousands more left by boat and vehicle, spreading the epidemic to every corner of the land. Nikolayev, not very robust and unable to withstand a Baku summer under normal circumstances, was persuaded by the Nobel doctor to join the exodus. Refineries and factories shut down, drilling stopped, stores were shuttered. Hagelin made his daily rounds with the regular company doctor and two additional physicians —a Russian and an American rushed to the area from Petersburg. A factory building was converted into a special barracks for the sick. Each morning the telegram went off to Emanuel listing the number of stricken employees and the totals for the city. Nobel losses were lighter than those of the other firms; a hundred fifty cases in Black Town with thirty fatalities, slightly higher totals in Balakhany.

In the epidemic's early stages Hagelin decided that the company should continue to operate as normally as possible. Alone among the producers Nobel never banked a distillery fire, never tied up ships or shut down pumps. He believed that remaining on the job was the surest means to battle the mental anguish of a plague, the insecurity of wondering if cholera would strike, the worry about friends and relatives; it was the only way to reach that stage of physical exhaustion which made it possible to sleep in the Baku evenings with temperature and humidity in the nineties. The summer was agonizing but by August the death rate started to decline and at the end of the month those who had fled began to return.

Elsewhere the situation was almost as critical for the land was twice cursed—these were "the hunger years." Cholera was accompanied by the twin scourge of famine. The correspondent of the *London Daily Chronicle* reported that "the bulk of the peasants have neither horses nor cattle to draw their ploughs. . . . without seed to sow their fields . . . there can be nothing before them but death and a slow and cruel one if not accelerated by disease."[4]

In Baku the stores gradually reopened, workers returned to the fields and factories, and the talk of the town again was oil. Emanuel's visit in October found Baku back to normal. He personally brought the gratitude of the board to Hagelin and his associates for the manner in which they had weathered another crisis. His confidence in the young Swede, his liking for the former pipe fitter, was the beginning of a close and lifelong friendship between Emanuel and Karl Vasilievich.

Hagelin had proven that he was technically qualified. He now set about to prove that he was emotionally and intellectually capable of taking command of so many colleagues and laborers who last knew him as an eager and ambitious apprentice. He was equal to the challenge. Calm, fair-minded, regarded as one of the most decisive executives in the firm, he soon demonstrated that he was as able an administrator as he was an inventive engineer. His talent for keeping his opinions to himself while patiently listening to the arguments of others was the subject of an outer-office anecdote: A visitor unable to get a reaction to a long presentation commented that Karl Vasilievich was being unusually silent that particular day, that he had not spoken more than two words. "On the contrary," replied Hagelin, "I said 'Good Morning' to you and now I am saying 'Good-bye'!"

His negotiations to purchase the tons of crude needed to supplement his own supplies—Nobel was by far the largest buyer of crude oil in Baku—were equally terse: a verbal agreement sealed by a handshake. He believed in an old Russian saying that written contracts only encourage both sides to figure some way to circumvent the commitment. Like the other Nobel leaders he was proud to be an exception to the commonly accepted dictum: "Whoever lives a year among the oil owners of Baku can never be a decent person again." As a man of his word he expected others to be equally honest, even in Baku. He was also a man of quick decision, impatient with those traveling a slower path, although Norwegian author Knut Hamsun, meeting him in his Baku office, found this "fine and amiable man" not at all impatient or in a hurry to usher a vagabond writer out of his busy life.[5] As Swedish-Norwegian consul for Baku and as deputy chairman of the Baku Oil Producers Society he needed patience.

The society was established in 1886 after two false starts during a month-long conference sponsored by the Ministry of Public Works, the Imperial Department of Mines, the military governor of Kutais commanding the port of Batum, the Baku branch of the Imperial Technical Society, and the individual oil producers. The conference and the permanent executive council it created were landmark developments in Russia, remarkable in the polyglot setting of Baku. Specific rights and responsibilities were delineated to the delegates by the railroads, steamship companies, and the oil industry; compromises were hammered out on voting and dues payment. Nobel and the largest concerns were given two votes and dues were levied on producers (.05 kopecks per pood), refiners (.03 kopecks), and pipeline carriers (.02 kopecks). During Hagelin's tenure these payments produced an annual income of well over a million rubles which the council dispersed to finance construction of schools, roads, a pharmacy, a hospital, and to pay technical and clerical staffs in the society's impressive offices who maintained statistical records, organized conferences, and published a fortnightly journal.

The society and its council represented the entire industry as well as national and local interests which had a stake in that industry. It was a forum for discussion, a deliberative body to settle disputes, to discuss issues vital to the trade: improvement of Batum harbor facilities; arrangement of credit advances from state banks; classification of oil cargoes; reduction of rail tariffs on chemicals used in refining; appointment of standing committees to allocate tankcars on the Transcaucasian railroad, to construct workers' housing and organize local amusements, to authorize and control trial drilling on new land. By 1914 its annual budget was close to 3 million rubles and the society was supporting nine different schools with fifty-seven branches, seventy-nine teachers, and twenty-four hundred students. Education of workers' children was free as were the four libraries. The Medical Aid Department maintained two hospitals with a staff of forty doctors and one hundred twenty-six nurses.

The Oil Producers Society was more representative of the total interests of the industry than any similar organization in the world. Its example and its success were directly responsible for providing precedents and creating a climate of opinion encouraging the sales and trade agreements that followed. J. D. Henry, the British expert who had studied the world's oil fields from Texas to the Carpathians, was thinking of these achievements when he wrote that nowhere had he found "a body of oil men who surpass in energy, enterprise, or business capacity those who are at the head of the industry in Russia."[6]

Ten years after the society's formation the government held its first public auction of the reserve lands surrounding Baku. Bids were based on a fixed cash royalty to be paid to the government for each pood produced. At later auctions in 1898 and 1899 the bidding averaged less than 3 kopecks a pood but by 1900 speculation was so rampant that the average bid for the forty-three acres on the block was just under 6 kopecks with some traders offering twice that amount. Hagelin was in charge of Nobel bidding and with Emanuel, the company geologists, and their local land agent Karasev, worked out the successful strategy which added substantially to Nobel holdings. By opening up the reserve lands they and the other successful bidders were guaranteed continued expansion of production. Rumors had spread that the Russian fields were drying up, the Baku wells nearing exhaustion. Professor Mendeleyev published a treatise— "On the Supposed Exhaustion of the Baku Oil Fields"—proving otherwise. The government's policy on reserve lands, the interest of the Producers Society in exploration of new fields, and the efforts of Nobel in leading that exploration meant continued supply at the record levels established in previous years. After 1900 when Balakhany production began to decline the slack was taken up by the newer Sabunchy and Romany fields to the east and by the extremely productive Bibi Eibat area south of town near the natural harbor of Bailov Point, out where the Caspian navy was moored.

A high percentage of the production still came from gushers. Eighteen ninety-five was a banner year with more than a million tons pouring from Romany spouters. The following year nine fountains in Bibi Eibat—where there were already two hundred twenty-eight productive wells—yielded another million tons. Significantly, none of these gushers came in at less than a thousand fifty feet. Nobel had pioneered the deeper drilling and the dramatic increases in Baku yields during these years was testimony to the merit of the innovation and the skills of the drilling crews.

During the dozen years of Hagelin's incumbency as Baku manager a total of one hundred thirty-eight gushers spouted nearly fifteen million tons of black gold. Such figures explain why Russia was producing more than half the world's petroleum. In 1901 the wells of Balakhany, Sabunchy, and Romany produced close to nine million tons and the entire peninsula almost eleven million, including the yield from Bibi Eibat, labeled "the most famous oil field in the world." Three miles south of White Town in a valley ringed by steep cliffs and bordering on the sea Bibi Eibat's forest of wooden towers by 1905 produced more than twenty-two million tons of oil from over two thousand wells.

There seemed to be no end to the supply but Hagelin, knowing that all good things do come to an end—even Baku oil—sent out his Swedish geologists and land agent Karasev to seek new sources. Karasev suggested they try two parcels of land the government had included in its original 1874 auction. They were in an area a hundred twenty miles north of Baku about twenty-five miles from Derbent, an isolated and deserted site called Berekey. Felix Hedman, an experienced drilling engineer, was put in charge of the first test probe. Drilling was relatively easy and in a short time he brought in a gusher. The quality of the oil was excellent, ideal for lubricants. A well was sunk on the other parcel of land and the results were the same. Baku soon buzzed with preparations for exploitation; plans were made for an electricity plant, pipelines, dock facilities, as Nobel and many other companies rushed to the wilds of Derbent.

The Berekey boom was on. In London, with a rising market for Russian petroleum shares, the interest was intense. British firms erected towers, sank shafts, and spent millions. Other companies joined the rush. But not another drop of oil was found. Nobel's two gushers, one yielding forty thousand tons and the other twenty-four thousand tons, were the only productive wells ever drilled. After two years they closed down with a profit. All the others lost heavily. The Berekey experience confirmed what oilmen have learned through the years but which was then new to the Russians: "Never believe in a gusher."

Closer to home Hagelin sent an exploring team out to Sviatoy, the Holy Island, a small piece of land in the Caspian where Count Witte's uncle once ran a small refinery. The results were not as sensational as those which set off the Berekey boom but the yield was good and the production steady for several years.

On the island of Cheleken across the Caspian on its eastern shore the results were far more impressive. An isolated spot inhabited by primitive Turkomans—before the Russians subdued the area their chief occupation had been piracy—the natives now supported themselves by fishing and the mining of ozocerite, a waxlike mixture of hydrocarbons. Cheleken had once attracted Robert Nobel. Included in his initial refinery purchase was a small section of land on the island where he drilled a pair of wells; then he gave up the effort to concentrate on his refinery, the wells and the island forgotten until Hagelin quite by chance learned of their existence and organized an expedition.

Cheleken oil was found to be rich in paraffin. Hagelin knew that Russia imported considerable quantities of this substance and he also knew that if Cheleken oil could be economically processed and the paraffin extracted the company could expand its product line and the country be spared the cost and trouble of importing at least part of its consumption. After examining the site with company geologist Torbern Fegraeus he returned to Baku and announced his decision. He ordered immediate preparations to commence exploitation on Cheleken, citing Peter the Great's warning to the Russian peasants that loss of time was the same as death. For Hagelin the time between the thought and the act was never long. A small power station was built to illuminate the area, men and machinery shipped out, casings pounded into the ground. In a few months a steady supply of oil came from the island and a steady stream of drillers and engineers from other companies were making their own explorations, producing their own oil.

Construction of a paraffin refinery was not so easily achieved. No one in Baku or Petersburg had any knowledge of the subject so Knut Malm was sent off to the Galician factories to learn their procedures. But when he returned, fully briefed and with a load of special equipment, he learned that Cheleken oil was of a different consistency and chemical content than the Galician and therefore had to be handled differently. Until he and the Nobel research team managed to find the right techniques the paraffin oil was sent up the Volga, mixed with the regular fuel oil, and successfully marketed. Once the plant was finished and production procedures mastered the new Nobel paraffin scored an instant hit on the market; the chief customers were the cloisters who had the state monopoly for manufacture of "wax" church candles.

Hagelin's decision to proceed with Cheleken exploitation, to explore Holy Island and Berekey and then as far afield as Dossor on the Ural river, were typical of his approach. If oil is to be found, go out and find it. If extraction is too expensive or quality too low, consider additional bidding on government reserve lands. Once in production, examine where costs can be cut and productivity increased. It was the kind of thinking needed in Baku. The expansion of the previous decade had been so rapid there scarcely had been time to make such analyses, to consider what reforms were in order, what kind of reorganization was needed.

He started with the subject he knew so well, transportation. A few years earlier it had been decided to reduce the speed on all Caspian tankers in order to save fuel. The idea might have had some validity in the beginning but when Hagelin examined the problem he learned that reduced speed and longer transport time forced the company to charter additional vessels to make up the loss; the cost of the charter was considerably more than the cost of the fuel saved by lowered speed. His statistics convinced the board. Three years after he put all tankers back on full throttle there was a 50-percent increase in the amount of oil transported by the same number of ships.

In the area of refinery operations Hagelin had to rely on the expertise of others, men like Axel Lambert whom he put in charge of all Baku refineries. With Olsen and his rejuvenated sales department selling more kerosene and lubricating oils in Russia and across the continent Lambert had to keep the quality high and the cost levels stable. In 1895 he commenced production of a new higher-grade kerosene, the Meteor brand; a company of that name was spun off from Nobel Brothers fifteen years later. By 1908 when Nobel refineries had a combined capacity of four hundred thirty thousand tons, a third of the Baku total, another new brand was introduced, Tunis, which had a higher flash point and was intended for the warmer climes of southern Italy and North Africa. These were the twilight years of the age of illumination but Nobel production and sales of kerosene were at record levels. It would be several years before more than a small fraction of the world enjoyed the benefits of electricity.

Lambert was also responsible for refining those "products which are worth their weight in gold," as Gulbenkian described them when writing with some horror of the usual Baku waste of vaseline, benzine, sulphur, varnish, and substances for aniline dyes.[7] And it was Lambert who endorsed the plans for a new sulphuric acid plant. The German chemist Dr. Niedenfuhr hired by Hagelin in 1893 made several basic improvements in Ludwig's plant and had started regenerating the acid residue but the sulphur still had to be imported, keeping the cost of the finished product high. The German Siemens firm which had been working a copper mine near Adjikent since the 1860s suggested to Hagelin that he buy their copper sulphide, extract the sulphur, and sell the raw copper back to Siemens. Hagelin enjoyed an excellent rapport with Siemens engineers; he had designed and built a pipeline and pumping station to carry oil to their mountain mine from the rail line below—Siemens had used up all the wood in the area and had been transporting oil on muleback. But the outlay for new plant construction to process copper sulphide was too great and Hagelin and his geologists had to look elsewhere for their supply of raw material.

The source was closer than they dared hope. Once more they confirmed the Georgian folksaying that when the Lord made the world he sprinkled all the minerals on the Caucasus.[8] Pyrites were discovered near Alexandropol (now Leninakan) and Nobel Brothers entered the mining business, putting up their own kilns. By 1906 the company was producing all-Russian sulphuric acid and useful by-products: hydrochloric acid, blue vitriol, and pure copper. Imported sulphur was replaced by a less expensive and more accessible local product. For the achievement Hagelin received a decoration from the government.

He showed similar foresight in evaluating the need for electrification of the oil fields, for providing electric power for the drills and pumps. Already responsible for the plants which introduced dockside lighting in Baku, Tsaritsyn, and other distribution points, the company did not hesitate to approve Hagelin's suggestion to construct a central generating station in Sabunchy. Other plants then were built to replace the steam generators in Bibi Eibat, Surakhany, Cheleken, and Holy Island.[9] Some were run on natural gas processed by a new Nobel Company gasworks. The largest unit was that built at Bibi Eibat where the company added a new administration building along with a workshop. The turbines were to be constructed by the Swedish STAL Company (Svenska Turbinfabriksaktiebolaget Ljungstroem) in which Emanuel was a major shareholder and held the rights for Russian fabrication and distribution, a logical diversification for the Petersburg plant. But war prevented the installation at Bibi Eibat just as it ended Emanuel's plans for producing and marketing still another major product.

Electricity was economical for drilling only if great quantities of its power could be generated; otherwise steam cost less. But only Nobel had the need and the capacity to make it economical so the other companies contracted with them for their supply. Or they signed with the municipal company, the Baku Electric Power Company—Bnito started buying from them in 1901, the year Baku Electric built a large station in White Town to service Balakhany and Sabunchy and another in Bailov for Bibi Eibat. Six-thousand-volt high-tension lines were strung through the fields and the electric motors near the wells were encased in fireproof brick sealed with iron doors. Such precautions were necessary to calm the fears of workers and owners. The latter opposed electric drilling because they could not regulate the velocity, but when they discovered that they could keep the rigs running at top speed and the workers matching it by running alongside they were converted to the new power.

Still larger storage facilities had to be built to store all the crude which the electric- and steam-powered drills were bringing from the ground. Not, Hagelin soon learned, those small pits for fuel oil he had supervised in the Volga district but mammoth reservoirs holding up to a hundred thousand tons. The largest one, with its circular shape and terraced stone lining, was quickly tabbed the Colosseum. Most of these pits were built between field and refinery; land was too expensive near the wells or close to Black Town. At Bejouk-Schor where apprentice Hagelin once had laid pipe from Balakhany a dozen new pits were constructed. The combined capacity was eight hundred thousand tons.

Gustaf Eklund, the veteran of the Petersburg factory, supervised the project and was then moved to the new fields purchased in the recent auctions. He organized the new work teams, built the power system to run drills and pumps, procured material for the towers, hired drilling crews, supervised the engineers, laid out barracks, planned the pipelines to the new Bejouk-Schor reservoirs. The towers were soon erected and the drills started their hammering, repeating a familiar pattern. Knut Hamsun described it: "The noise is tremendous. Dark Tatars and yellow Persians stand by their machines. . . . a tool goes down into the earth and returns after 50 seconds with 1,200 pounds of oil, goes down again, is gone 50 seconds and comes back with 1,200 pounds of oil. Around the clock, all the time." Lest the reader of his tales from the Caucasus, *In Wonderland*, think that this was like coining money from the lower depths Hamsun was quick to add, "The well cost money, it is 500 meters deep, it took a year to drill and cost 60,000 rubles."

The new field organized by Eklund marked the beginning of a new era for Hagelin and the company. With the Bejouk-Schor storage capacity Nobel Brothers could always be in command of more crude than all of their competitors and could for the first time set the price in the marketplace. More efficient refining techniques and less expensive transportation placed Nobel, by the late 1890s, in position to control, to regulate, and not just to dominate the domestic market. The company could exploit favorable market conditions, could take advantage of its tremendous storage and production capacity as never before.

Hagelin was quick to seize the opportunity. The Persian Schamsy Assadouleff, a moderately successful producer and refiner, brought in a gusher in a new area at the edge of the Romany field. He could not cap or control the well and sixteen thousand tons a day streamed into a nearby lake—hardly a suitable reservoir for volatile, flammable oil. He had neither pipeline nor adequate storage facilities. On the fourth day of the uncontrolled spouting he stopped Hagelin just outside Villa Petrolea as the Swede was driving his 5-horsepower Benz to the office. En route the pair concluded the largest purchase of crude oil in Baku's history. The first sixteen thousand tons were bought at the current market price, the remainder on a declining scale; but the deal depended on Hagelin's guarantee that within three days he would be ready to take a minimum of eight thousand tons every twenty-four hours. Wanting to get the oil into his reservoirs as fast as it was coming from the ground, in three days Hagelin was ready to take twice as much. At the end of five weeks Bejouk-Schor had a half-million tons of crude, Assadouleff was a millionaire, and Nobel Brothers had earned its entire year-end dividend. The amount paid for the oil was just half the going market price. It was a fast, painless profit.[10]

There were no contracts, no papers, no lawyers, just a verbal agreement and a handshake. When Hagelin asked the Persian to send over some workmen to check the amount of oil coming into the Nobel pits he replied that it would not be necessary, that he trusted the Swede. Each day Karl Vasilievich went to check the tanks, to note the day's flow, and then to send the payment to Assadouleff. This in a city condemned for its dishonesty, for what one journalist dismissed as "fierce financial competition, mad speculation, colossal frauds, thirst for monetary gain."[11]

The board in Petersburg had the same trust. How could they argue with Hagelin's performance? Providing the equivalent of the company's dividend on one transaction would have been proof enough. They had learned to respect his judgment, to listen to his suggestions. When he proposed establishing a sales network in Central Asia they posed no objections.

The commerce of the Caucasus was firmly in the hands of the Armenians who made their own deals with Armenian oil producers, but Central Asia was too far afield for them. Petersburg was occupied with other areas of Russia and Europe and had no time to worry about opening a new and small market in the east so they accepted Hagelin's proposal to send an old schoolmate from Saratov to open the territory. N. F. Skvortsov, for several years the company's bookkeeper in Tsaritsyn, was a wise choice. With Hagelin's guidance he organized the purchase in Krasno-vodsk of a small storage and distribution installation established previous-ly by one of the smaller independents. Skvortsov immediately modernized and enlarged the buildings, arranged for Nobel tankcar deliveries, put up a workshop and a small plant for fabrication of steel barrels; the usual wooden containers were not solid enough for the rigors of Asian rail traffic. Those of steel were expensive to make and 25 percent more expensive to ship but they were a necessary evil and the cost was passed on to the con-sumer. Skvortsov prospered and in a few years he spread his sales territory to the northeast corner of Persia erecting a small factory in the pilgrim city of Meched, southern terminus of the Transcaspian railroad and departure point for caravans making the trek to the Persian heartlands. Nobel oil was carried on camelback across the sands of what one day would be exploited as the greatest oil field in the world.

Central Asia became an important sales area but Hagelin could not continue to monitor the progress of his old friend. In 1899 he was sum-moned to Petersburg and appointed to replace the aged Beliamin as one of the five company directors. He was placed in charge of the Baku operations and the entire transportation network. After twenty years in the company he had reached the summit. He was one of the five who directed the destinies of the Nobel empire. On the twenty-fifth anniversary of that empire Nobel Brothers had over twelve thousand employees and a payroll exceeding $2.5 million. There were thirty-eight hundred workers in the oil fields, thirty-six hundred in the machine shops, the five refineries, and a half-dozen auxiliary factories, a hundred in the Baku offices, an equal number in Petersburg, a string of workers in the hundred fifty depots and tank farms, twenty-three hundred in the Nobel fleet including two hun-dred fifty officers. But they could have used more officers and more vessels; the company was paying $5.5 million in freight and chartering charges to shippers and almost as much to the railroads.

"We had in our service," Hagelin wrote, "Swedes, Danes, Norwegians, Finns, Germans, Balts, Russians, Armenians, Persians, Georgians and even a Frenchman, but all these different nationals were first and foremost Nobelites."[12] The Nobel Company was their company, the successes and failures their own successes and failures. There was a pride in belonging to the pioneers, to the best, a loyalty and special spirit in being identified as a Nobelite. In the coming years that loyalty and spirit would be severely tested.

CHAPTER TWELVE

Revolution on the Caspian

Swedish or Russian, Armenian or Tatar, the Nobelites were still subject to the will and whims of the tsar and the most reactionary government in Europe. Within their own empire the Nobelites were still dependent on the empire of the Romanovs and ruled by it. Nowhere in that empire were the weaknesses, the sins of commission and omission more glaringly exposed than in the Caucasus. The respite and relief provided by the enlightened administration of a Prince Vorontsov were infrequent occurrences, aberrations in the routine bureaucratic reliance of granting promotion to volunteers, of using the area as dumping ground for deadwood. Careerists and opportunists joined the merely incompetent in providing the ideal circle of courtiers for Petersburg's next idiocy, the appointment of Prince Golitsyn. A tsarlet of the narrowest kind with neither brains nor background, his was the talent most compatible with those who served him. The policeman's baton was his only instrument of government as he tried to hammer through the policies of Russification while tolerating the socialist agitators and other troublemakers who had been exiled from the north for safekeeping in the Caucasus—an official policy as sensible as many others from the witless Nicholas II.

Golitsyn and his government were far less concerned with industrial unrest in Baku, Batum, and Tiflis than they were with the threat of Georgian, Tatar, and Armenian nationalism. In many instances the tsarist secret police encouraged the socialist agitators, promoted their strikes and demands for better conditions, basing their strategy on the hope of diverting attention from what they—in all their wisdom and frequent overindulgence in the subtleties of countersubversive tactics—believed to be the greater threat. The regime insisted on Russification, on autocracy, orthodoxy, and nationalism as dictated and interpreted by Grand Inquisitor Constantine Pobedonostsev who was preaching intolerance across the length and breadth of the land, attacking the Finns in the north, the Poles in the west, the Armenians and Georgians in the south, and the Jews wherever they were found. Baron Alphonse, the septuagenarian head of the Paris Rothschilds, complained to Witte about the court's preoccupation with the mystical and the occult which so often translated into pogroms, mass expulsions, and the most vicious form of discrimination against the Jews.[1]

Interior Minister von Plehve, blackest of the reactionaries and organizer of those pogroms, moved as deliberately against the Armenians as he did against the Jews. He ordered the confiscation of the property of their national church. Plehve and his will-less tsar in a single stroke destroyed the loyalty of a group that had been among His Majesty's most loyal subjects. After the horrors of the Turkish massacre, when the sultan declared that the only solution to the Armenian problem was to do away with the Armenians and the Russian government did nothing to halt or moderate the slaughter, the government suppressed the autonomous Armenian schools. Seizure of church properties was the final and telling blow. The apostles of the new social democracy swept into the Caucasus preaching reform and overthrow of the old system and the Armenians were among the most enthusiastic supporters of the new gospel.

The most fertile and receptive soil for spreading that gospel was found in the towns where the industries were concentrated, in Baku, Tiflis, and Batum, in the oil fields, the workers' barracks, the tobacco and barrel factories, the railroad yards. The very nature of Russian industrial development with its many gigantic enterprises provided an ideal setting for agitation and the preaching of reform. Only the army and navy brought more people together in concentrated groups. A third of Russian factories employed more than five hundred workers; in Petersburg over 70 percent of the factories had that large a payroll and by 1900 Georgia had fifty thousand industrial workers. With their families they accounted for 10 percent of the population—and it was a concentrated 10 percent. Tiflis alone, a city of two hundred thousand, had in excess of twenty-five thousand workers and a few thousand more were in the railway workshops and warehouses. There were at least that many in Baku which had nearly trebled its population from the forty-five thousand of the mid-eighties.[2]

Packed together in factories and inadequate housing, the workers often lived in primitive conditions of indescribable filth with insufficient ventilation, heat, or light, complete lack of safety considerations and absolutely no compensation for accidents, no bathrooms or canteens, no unions or the right to strike or even to protest. There was compulsory over-time—often without compensation—enforced purchasing at company commissaries, payment in kind. All this to earn 50 cents a day for twelve hours' labor. In Batum the average was fourteen hours with another two of compulsory overtime. Many of the eleven thousand workers, a fifth of the population, made less than 50 cents a day as they strained in the barrel and case factories, the tobacco workshops, iron foundry, bottling plant, or along the busy wharves.

The first rumblings of real discontent occurred in Tiflis in 1898–1899 with small and easily suppressed strikes at the railroad yards, the tobacco plant, and the shoe factory where Stalin's father had once worked and where conditions were particularly primitive. Echoes were heard in Batum at the Rothschild installations. The young Stalin, Iosif Dyugashvili, a student at Tiflis Theological Seminary, was spellbound by the tales of political exile and Siberian sufferings relayed by escaped revolutionaries as he tutored the workers in the rigors of Marxist dialectic and helped instigate another unsuccessful strike in August 1900. By then, with record oil production and a glutted world market, Baku's industry was severely depressed. There was also an oversupply of manganese and that commodity dropped in price drastically as did oil. Payrolls were cut back and the area suffered from a general economic decline as Russia reeled from the impact of the American and European depressions. There was another railroad strike in Tiflis, this one more serious than the first and organized by one of those exiled troublemakers from the north, Mikhail Kalinin, future president of the Soviet Union. A poor harvest in 1901 added to the troubles, and when police fired on a May Day crowd in Tiflis that year Lenin characterized the event as "the beginnings of an open revolutionary movement in the Caucasus." From then until the final debacle in 1917–1918 the Caucasus showed the way for the rest of the revolutionaries and still another title was added to the notoriety of Baku. It had the distinction of being "the revolutionary hotbed on the Caspian."[3]

The world would not know how hot it really was for many years, but the credits then were confused as Soviet historians made everyone else shrink into Stalin's shadow. Sycophantic idolators like secret police chief Lavrenti Beria—whose Bolshevik career commenced in Baku in 1917 while he was a student at the Technical College—in his official history of the Bolshevik Party in the Caucasus placed his leader at the head of every demonstration, making every important decision, proving his undeniable genius in every conceivable way. In truth Stalin played a relatively minor role although Beria credited him with leadership of one of the most remarkable operations in the history of underground political warfare.

In the heart of the Tatar quarter, past dozens of small shops where filagree silver, carpets, and ancient Persian weapons were sold in the shadow of the khan's palace and along the winding alleys with their flat-roofed windowless houses, there was a large cellar which ran under several buildings. Here worked as many as ten dedicated party functionaries operating printing presses, binders, cutters, setting type in several languages. This was "Nina", headquarters for printed propaganda for Baku and the Caucasus, then for southern Russia and eventually the entire country. Ten thousand copies of Lenin's *Iskra* were published in the Baku basement. Lenin's wife made up the mats, shipped them concealed in bookbindings to a local front man via Persia, and printers used them to make castings. The paper first appeared in 1900 printed in Stuttgart and later in Munich, Geneva, London, but always reprinted and distributed from the Baku basement. The Okhrana could not understand how so many copies of *Iskra* kept entering the country; they sealed the European border, then the Persian—still the paper was found in all corners of the empire. So were copies of the "Communist Manifesto," the illegal Georgian Marxist publication *Brdzola*, pamphlets by Lenin and Leon Trotsky. The handful of men in a stuffy cellar working ten to twelve hours a day for $12 to $15 a month printed more than a million pieces of propaganda.

From the Tatar quarter of Baku the printed word was distributed in a manner making the selection of this area as wise a geographic choice as it had been incomparable for its clandestinity. From Baku a steady stream of rail traffic went west to Tiflis, Batum, and Odessa, north to Petrovsk and Moscow; hundreds of tankers and other ships sailed to Astrakhan and the Volga which led into the heart of the country. It was not difficult to recruit couriers, to arrange for concealment, disguise contents, and falsely label packages. The Nobel distribution network was a perfect channel for dissemination and must have been a prime target of the conspirators organizing the spread of their propaganda.

The most important support "Nina" received during its early years came from a truly remarkable individual who would later become Minister of Supply for Trotsky's Red Army, chief Soviet diplomat and negotiator, Commissar of Trade and Industry—Leonid Borisovich Krassin. Too long mired in the sludge of Soviet historiography and misdirected hagiolatry, his achievements should have ranked him in the highest circle of Communist saints and heroes.[4] He was born and raised in the western steppes of Siberia and attended Petersburg Technical Institute until his expulsion for socialist activities. After a short period of exile working as a draftsman on the Siberian railroad he was permitted to complete his education and—banned from northern cities—in 1900 headed for Baku where a friend procured him employment building a second power station in White Town. A respected engineer with considerable charm and an attractive, engaging manner, he was on excellent terms with oil company executives and engineers and was soon promoted to manager of the Baku Electric Company. He used the office—a haven for socialists from south and north—to arrange for smuggling supplies into the clandestine printing plant, for shipping papers and pamphlets across the country. Passports were forged, refugees from the Okhrana secreted in and out of the city. He was a tireless and highly successful fund-raiser, organized party affairs in an efficient and businesslike manner, at one point even procured a loan of 2,000 rubles from the municipality ostensibly for power station expenses but actually to purchase equipment for "Nina."

An extraordinary man. Lenin called him the Bolsheviks' Minister of Finance. Thin and bony, never in robust health but a bit of a ladies' man, described by friend and admirer Maxim Gorky as shrewd-looking with the face of an old icon, Krassin stayed in Baku four years. He departed only when striking workers in the power plant demanded his dismissal as a reactionary enemy of the working class. For a time he worked with a Belgian firm in Moscow and then Petersburg where he played an active role preparing and leading the uprisings of 1905. In 1908 he went to Berlin to join the Siemens Company. Four years later he was back in the Russian capital as managing director of the local Siemens office, serving on the board of several other companies as well.

His was a model double life: respected in the business community yet an effective underground leader entrusted with responsibility for illegal fund-raising, including robbery and terrorism. While serving as the Petersburg businessman of distinction he made several trips to Tiflis and Baku to guide those in the area where the art of socialist extortion had proven particularly profitable. Other socialists leaned on him for advice and guidance. His stewardship of the "revolutionary hotbed on the Caspian" had taught him the means of survival, and hardened him for the long and difficult years ahead, just as other Caucasian apprenticeships were preparing Stalin, Kalinin, Klementi Voroshilov—secretary of the

Baku Oil Workers Union and future Soviet marshall—and so many others who would emerge as leaders of the new Soviet state.

To replace the valuable Krassin in Baku Lenin sent out Lev Kamenev and Vladimir Bobrovsky whose wife was secretary of the Baku committee. She reorganized the clandestine press and aided Kamenev in his vigorous efforts to rally the emerging Bolshevik faction in the face of socialist opposition, the Menshevik majority whose stronghold in all of Russia was centered in the Caucasus. Kamenev, married to Trotsky's sister and another of the northern troublemakers exiled to the provinces, founded the Transcaucasian Bolshevik Party in Tiflis in 1904. In Baku leadership of that party was assumed by Stepan Shaumyan, friend of Plekhanov and Lenin, delegate to the important London and Stockholm party congresses of 1906-1907. He was chief organizer and propagandist, editor of legal and illegal publications, and in 1917 the logical choice to be chairman of the Workers' Soviet of Baku; after the revolution he was chairman of the Baku Council of Peoples' Commissars.

The oily sands of Baku with all the problems and abuses of a raw and powerful frontier capitalism at its laissez-faire worst were fertile soil to these revolutionaries. As the stumbling, bungling tsarist government compounded its errors in efforts to govern the area, to put down the trouble and suppress the dissatisfaction, the socialists' cause was further strengthened. The area became the training ground where the lessons of strike and protest, the organization of the proletariat, the enforcement of clandestineness were learned by those who would lead the Soviet Union. In what Stalin termed "the storm of the deepest conflicts between workers and oil industrialists" he received his "revolutionary baptism in combat" and for the first time "learned what it meant to lead great masses of workers." He later stated that "the revolutionary work among the oil workers of Baku hardened me as a practical fighter and as one of the practical leaders."[5]

Stalin's baptism in Batum came during the 1901-1902 strikes, protest parades, and demonstrations against the factories of Mantashev and Rothschild and the repressive policies of the military governor whose order to fire on the demonstrators—slaying and wounding the workers—rallied those moderates who had not supported the strikers or the tactics of Stalin. The police cracked down; Stalin and many others were arrested. It was the first of his eight arrests as he repeatedly escaped from exile and repeatedly found himself back in Baku's Bailov prison.

In July 1903 a strike was called by the oil workers in Baku. It was a spark that set off flames which spread like a brushfire across southern Russia as workers in Batum, Tiflis, Kiev, Ekaterinoslav, Odessa, walked off the job. It was the country's first general strike and caused a governmental crisis, the dismissal of Count Witte, and the emergence of Plehve as the supreme power of the realm. That minister's long hoped-for "small victorious war"—intended as a distraction from the results of the repressive policies of the government, a means of rallying the nation—ended in unmitigated disaster as the Japanese destroyed the Russian fleet, humiliated the army, and placed the government in jeopardy. Plehve was removed from the scene of his crimes by assassination at about the same time as his partner in reaction and misdirected incompetence Prince Golitsyn was departing his post in the Caucasus. But the pair were around long enough to witness the effects of another catastrophic policy of iron-man Plehve as ill-conceived as his little war but of greater harm to the country he served so poorly. That was the scheme of the Black Hundreds, another attempt to divert the attention of the workers from their economic misery, to divert the masses from thoughts of their immediate problems by providing scapegoats, to encourage Tatars to turn against Armenians and loyal Russians everywhere against the Jews as the evil exploiters responsible for the troubles of the nation.

This vicious and logical extension of the policies of orthodoxy, autocracy, and nationalism was a vigorous and fitting climax to the years of official anti-Semitism and suppression of nationalities within the empire. The instruments of that policy, the Black Hundreds, were never better described than by Bertram Wolfe who characterized this "most unholy and variegated band" as:

> the backward, the degenerate, the brutalized, the bewildered, the enraged, the entrenched, the ruined; officers, landowners and gilded youth, demobilized old soldiers and personal servingmen to whom loyalty to the master was the sum of all loyalties, criminals whose police records made them amenable to any instructions, ruined artisans and shopkeepers who were persuaded that the strikes and the eight-hour day were the cause of all their woes, hungry, degraded slum proletarians from the human scrap heap of the great cities, the more illiterate and credulous among the workingmen and peasantry.[6]

They swept over the troubled land like the four horsemen of the apocalypse spreading death and destruction in their path, reaping a wild wind of hatred and religious fanaticism. Exploiting the demands of the workers, inciting Tatar against Armenian and Armenian against Tatar, attacking Jew and foreigner, they fired the oil fields, stormed the refineries, vandalized the offices, invaded workshop and home killing anyone who tried to put a stop to their depravity.

The evil of the Black Hundreds descended on Baku as the city was slowly recovering from the troubles of 1904 when the entire countryside erupted into what Lenin believed to be the Grand Rehearsal. In December there was a general strike of the oil workers, the first large-scale strike in Russian history that did not begin spontaneously—it was planned and organized by the Mensheviks who for the first time were testing their ideas about the use of mass strikes in bringing about revolution.[7] The workers were still on strike when news of Petersburg's Bloody Sunday reached Baku. Tales of the bloodstained snow, the senseless slaughter of Father Gapon's petitioners, were slow to reach the Caspian and Baku had been cut off from the rest of the country for two weeks by an unusually severe snowstorm. Many froze to death in the Tatar quarter. When the town finally learned that the tsar's troops had fired on defenseless workers there was another general strike which soon spread to Batum, Tiflis, and throughout the Caucasus. Gapon's mournful words after the shooting—"There is no Tsar anymore"—recalled the old proverb: "Without the Tsar the land is a widow, the people an orphan."

Brutal murders and bold seizures of land, open dissemination of revolutionary propaganda, marching protesters behind flowing red banners, workers' councils with a flood of demands, a complete halt of trade and commerce resulted. The battleship *Potemkin* was seized in Odessa and sailed across the Black Sea, the officers and defenders of the regime put to death. There was another general strike in June but by late summer the local police and Cossacks commenced their own revolts, swiftly suppressing the workers and their leaders. By the end of the year, as a calm settled over certain sections of the country, many cities and villages in the Caucasus were without any central governmental authority or restraint. They were making their own decisions completely independent of Petersburg.

In Baku there was no calm, no peace. The Tatars demanded arms from the governor in order to defend themselves against the Armenians whom they claimed were plotting to seize the town. Traditionally it was a heavily garrisoned area but levies for the war with Japan had reduced the military presence and when the Armenians, converted by government stupidities into implacable foes of the regime, assassinated a dozen local officials the Tatars were given weapons. The dozen were more than avenged. As local troops stood by Armenians were slain in the streets or besieged in their homes. The correspondent of the *London Daily Mail* reported that it was "one great outburst of vengeance and hatred."[8]

One of the wealthiest Armenian oil barons withstood a siege in his oriental palace of a home for three straight days. His brother had been murdered and mutilated in the Baku park a few months earlier and he knew the price of surrender. With his Winchester repeater he picked off forty of the enraged Tatars before they managed to set fire to the house and overwhelm the sharpshooter. Members of his family and staff found alive in the cellar were murdered and dismembered before a shrieking crowd. A similar fate befell another wealthy Armenian. Responding to the pleas for help the governor had promised immediate assistance and protection, but his forces did not arrive until the house was burned out and all members of the family massacred. For three days the governor did nothing. Then the religious leaders, the bishop and the mullah, marched together through the streets declaring a truce. Tatar and Armenian alike carried out their dead, laid them against the walls, and embraced one another.[9]

As if that were not enough blood for the harassed inhabitants of Baku, for the first time the government allowed the Shiite Moslems to hold their annual "Shakhssey-Vakhssey" ceremony honoring the death of Mohammed's "rightful heir" in the center of the town. Previously restricted to Kishli, a lawless and crime-ridden Tatar village between White Town and Balakhany, the black-and-red-robed Tatar penitents paraded for ten days through the streets of Baku, mournfully keening, beating chests and backs with chains, moving slowly to the rhythm of the drums. On the tenth day a group of white-robed men appeared, sobbing and screaming and climaxing their ceremony of flagellation with the bloodiest act of all: self-mutilation. Razor-sharp swords, kingals, were given to each man who slid the blade slowly across the top of his shaven head. The white robes turned crimson as some of the penitents were carried off while others, wild with the frenzy of the moment, had to be restrained from further mutilation.

An awesome and symbolic sight in this period of insecurity, savagery, and brutality defying description as the streets of the city ran red with the blood of Tatar and Armenian, inscribing the darkest chapters in Baku's rich and varied history. The mayor's claim that "they themselves do not know why they are killing each other" was disputed by both groups and at the end of summer the death toll reached two thousand. The killings stopped. But it was only a truce and there were cries for revenge. More than half the Baku workers were Tatars, Persians, Lezgins, or Turks, but the several thousand Persian laborers mainly were migrants from across the border; they worked half a year and usually were eager to return to their homes. Neither they nor the Turks instigated the slaughter. It was the Tatars, the poorest of the poor, the natives who had seen the Armenians gain economic ascendancy in the area they once dominated. It was the Tatars who sought revenge.

By the time the slaughter stopped, Witte—with the aid of the American president Roosevelt who won a Nobel Peace Prize for his efforts—had negotiated a peace treaty with Japan. The government felt secure enough to commence full-scale repression of the revolutionaries but progress was slow, especially in the Caucasus. In Tiflis the local soviet had taken command of the workers' section of town and in Batum the workers had erected barricades to defend themselves against marauding Cossacks. In Baku the volcano again erupted on 2 September just as news of the humiliating peace treaty was reaching Russia.

This time the Black Hundreds stormed into the city, firing the oil fields, openly putting to the torch hundreds of derricks and reservoirs. The wooden towers crashed down in flames and oil tanks exploded amid the crackling of rifle fire; smoke and soot blackened the sky and shut out the sun. It was a scene from the last days of Pompeii, a living inferno where for a week thousands of Tatars again ran wild, searching every building for stray Armenians, shooting at burning men fleeing the flames, wielding their kingals with savage butchery, expertly mutilating the corpses. The Armenian manager of the Khatissov Company was found hiding in the cellar of a Nobel building and was murdered; another manager, a Frenchman with the misfortune of looking like an Armenian, was stabbed to death as he fled from a burning factory. Foreigners were not the targets but the British took no chances; they chartered a ship and rode out the storm in the safety of the outer Baku harbor. The targets were Mantashev and the other Armenians. Tatars and Black Hundreds fired their properties, destroying the installations of the Russian Oil Company and the Baku-Russian Petroleum Company. Just to make sure there would be no way to fight the blaze they also wrecked the local waterworks.[10]

Thousands were slaughtered as the rampaging mobs pillaged and looted. Cattle and horses were led off to Tatar enclosures; the abominable pigs were put to death as quickly and surely as the Armenians; whatever could not be carried away—furniture and other valuables—was burned. In less than a month's time the ravaging hordes destroyed over a thousand wells. Bibi Eibat was in a less vulnerable location and did not suffer the almost total destruction of Balakhany, but this "most famous oil field in the world" lost more than three hundred derricks and a hundred producing wells were set ablaze. It was as though all the forces of evil in the land had been let loose to run riot in obedience to the meanest of man's instincts—as though the devil himself were directing the destruction, obeying that proclamation by Stalin in the underground press: "Workers of the Caucasus, the hour of revenge has struck."[11]

Damage reached the millions. The industry was completely crippled. Never again would Russia lead the world in oil production. Its annual rate of nearly eleven million tons during the first years of the century would not be reached again until the late 1920s.

Nobel properties were not seriously damaged. Along with Schibaieff, Anglo-Russian, and the Russian Petroleum and Liquid Fuel Company they were spared the full fury of the hordes. Although acknowledged to be a foreign enterprise it was also Russian; Emanuel was a Russian citizen and his company's installations in Baku, Petersburg, and from one corner of the empire to the other were not targets of the regime or of its unofficial executioners, the Black Hundreds. But the Jews were marked and Rothschild properties in Baku and Batum—the Bnito and Mazut installations—were attacked. When the government learned of the Rothschild loan to the Japanese those attacks were intensified.

Nobel Brothers did not escape the Grand Rehearsal without some scars. Its obdurate stand during the 1903 strike when it refused to recognize the right of unions to compel membership of all the workers, to allow unions the power of decision on matters of workers' welfare and compensation—Nobel alone held to these positions while the other firms, fearing retaliation and not as confident of their workers' loyalty, gave in after the first weeks of unrest and the first serious strikes.

Hagelin was respected by the labor spokesmen, most of whom he had known for years; as the company representative he stubbornly refused to surrender. He entered the negotiations after the first company representative to the fray Gustaf Eklund—a forthright, authoritarian type and chief of the Baku complex—narrowly missed assassination. He escaped with his life but several of his associates were seriously wounded. Karl Vasilievich arrived and sent Eklund to Petersburg. He also took the precaution of arming himself while he made his rounds, bringing peace to the area and placating the workers while refusing to recognize their demands. When Baku returned to normal he went back to the main office and Artur Lessner was appointed to the post of chief in Baku. He had been director of his father's factory in Petersburg but when the banks started meddling in his affairs he resisted the interference and transferred to Nobel Brothers. His views and actions were more sympathetic to the workers than Eklund's had been and the young Lessner proved to be an unusually able leader of the multifaceted complex in Baku.

Lessner's troubles with the banks might have been manageable by Emanuel, by now a familiar figure in the financial institutions of Russia and Europe. He was a frequent borrower but his credit was established, his reputation for solidity made, and the future of the company bright. His uncle would have been proud of the achievement. The 1905 revolution—the massacres and pillage and destruction in Baku—of course affected that positive assessment of the company, but Emanuel had already made a substantial loan to tide him over the transition period and to cover any abrupt change in the overall health of the industry. As news of Bloody Sunday was appearing in the European press he was completing the preliminary negotiations at the Berlin Discount Bank for a twenty-year 5-percent loan of 32.4 million marks.

With so much unrest in the first years of the century raising money or issuing new stock was not too easy in Russia, but the German bankers were not bothered by a few industrial disputes. Nor were they particularly concerned when they learned—at the very moment when Emanuel was in Berlin to sign the note—that Grand Duke Sergei, governor general of Moscow, had been assassinated. They knew their company and they knew Emanuel. In 1897 they had loaned him 21 million marks. This newest issue was planned for release to the public on 25 February but on the 24th the news of the first Baku murders broke in Berlin and there were reports and rumors of anarchy, burning oil fields, vast destruction of property. The bank delayed circulation of the loan.

That was no longer Emanuel's problem. The note had been signed; it was the bank's job and not Nobel's to float it to the public. Emanuel had troubles of his own. Hagelin was reporting on the workers' demands; Lessner was reporting continued unrest. During September's holocaust Emanuel must have wondered about the future of the company, the future of the country itself. The burning oil fields of Baku could be smelled in Moscow and Petersburg where they had their own fires of revolution. The general strike spread to the capital, across to Finland; propaganda was distributed openly on the streets; there were demands for overthrow of the government. Nobel offices and the factory were shut down. There were no newspapers, no trains. The banks and all other businesses and shops shuttered their doors and windows. Everywhere were meetings, hand-bills, threats, and street-corner orators. Olsen could not believe what was happening. He took to his bed for weeks, his nerves shattered. He blamed it on what he called Russia's original sin, the lack of order and organization.

The newest of the Nobels, George Oleinikoff, a Russian doctor who had recently married Ludwig's daughter Marta, could not accept an inter-pretation so Germanic in its philosophy, so simple in its dismissal of legitimate grievances. He sympathized with the workers' demands and was arrested while attending a meeting of the Social Democrats. The world press rocked with the sensation of a Nobel in a tsarist prison. Despite Emanuel's strongest representations and all his connections in the government he could not gain a release for two months. But a two-month term and release for lack of evidence was a mild punishment at a time when the government was exiling thousands to Siberia and the Cossaks were given a free hand to crush the demonstrators.

Nicholas dissolved the first Duma and with it the hopes of so many who saw the first glimmerings of parliamentary democracy. He appointed Peter Stolypin as prime minister and "Stolypin's neckties" soon were be-ing erected in cities and villages for prompt execution of revolutionaries. The ruthless repression had an effect. The country was returning to nor-mal. But not the Caucasus. At the end of the year 1906 Nobel's chief in Batum was murdered on the street; a few days later the British consul and general manager of Schibaieff, Leslie Urquhart, was pulled from his automobile and beaten, narrowly escaping with his life. The next year Baku was swept again by a revolutionary ferment, by a strike which for a time threatened to become as general as those of 1904 and 1905. But the fires had burned out; these were a few flashes from the smoldering ashes of what had gone before.

By then the tsar had dissolved the second Duma and his ministers had revamped the election laws to ensure that the opposition groups—the Social Democrats, Socialist Revolutionaries, and the Cadets—would re-main in the minority. The Baku clandestine press continued to attack and the organization running "Nina" continued to display a vitality unique in the nation. The oil workers elected Bolshevik deputies to the Duma. They kept up a steady agitation for inclusion of the workers in one large industrial union rather than the smaller trade or craft unions and before the last flames died out they succeeded in gaining recognition of the right of the oil workers' union to negotiate directly with the representatives of the industry. It was a victory for the socialists, for the workers, for the right of collective bargaining. A British observer on the scene reported rather imperiously that "labor has been meddled with, unsettled and made offensively dictatorial, and a splendid industry has been placed in jeopardy,"[12] but by 1910 that meddling had brought the oil worker benefits previously reserved for the Nobelites: an eight-hour day in the fields, nine in refineries and factories; housing for nearly three-fourths of the work force with over 20 percent of those not housed receiving compensatory allowances.

The rest of the country was being crushed back into an apathetic acceptance of autocracy, orthodoxy, and nationalism, but the "revolutionary hotbed on the Caspian" was experiencing the nation's first significant show of proletarian power. From his distant exile in the west Lenin praised the Bolsheviks of Baku as "our last Mohicans of the political mass strike."[13]

CHAPTER THIRTEEN

Another Nobel Enterprise

There were no Mohicans for Lenin to praise in Petersburg or any other cities in the north. As the waves of revolutionary fervor receded a stranded, diluted Marxism was left behind, a discouraged and frustrated band of reformers. The spirit of dedication and sacrifice succumbed to self-doubt and disintegration of determination as those who escaped prison or "Stolypin's Necktie" scurried for safety beyond the clutches of the Okhrana.

The return to routine in the Ludwig Nobel factory was immediate and without incident. Although it was located on the fringe of the great workers' quarter of Petersburg the plant suffered no vandalism or violence. The real meaning of revolution would not be known in the north for another ten years. In 1905 Nobel workers did not seek revenge, did not turn on their employer and their shop foremen—although they did vote to create a permanent workers' committee "to eliminate misunderstandings between the management of the plant and the workers" as well as deciding on "pay rates for new articles of production." This precedent-shattering decision was made during the course of a series of meetings inspired by the government's action three weeks after Bloody Sunday: Petersburg workers were directed to elect representatives who would participate in a commission empowered to determine the causes of workers' discontent and to explore means of eliminating those causes. Representatives of the metalworkers—nearly half of the electors—met at the Nobel plant.[1]

Emanuel was an enlightened entrepreneur but he was not willing to relinquish much power of decision over production and personnel policies in his factory. He was confident that the Nobelites in Petersburg, like the Nobelites in Baku and along the Volga or anywhere else in the empire, were far better treated than their colleagues in other companies. For Nobel the welfare of the workers was still as primary a principle as it had been during the days of Ludwig. In the 1896 Nizhni-Novgorod exhibit the factory had received its third Imperial Eagle, this one awarded in recognition of the company's "concern and solicitude" for its workers. Even in the midst of revolution and social unrest the Nobel organization was still far ahead of its capitalist colleagues.

Emanuel built additional units at the workers' housing complex on Sampsonievsky and constructed apartments for salaried personnel. He provided free medical care for all workers—the doctor in charge was the Nobel family physician, his daughter married to Emanuel's brother Carl. A savings bank was established along with a special workers' emergency support fund. Ludwig's free educational opportunities now were broadened to include a school for the workers' children. On a large parcel of land Emanuel had purchased across Sampsonievsky Prospekt, stretching all the way to Lyesny Prospekt, he put up the new school building as well as housing units for factory engineers and other employees; there were also garages and a carpenters' shop where wooden models for castings were assembled.

Among the smaller private charities Hans Olsen's wife "adopted" twenty workers' babies each year giving them sterilized fresh milk, arranging for medical care and instruction in proper care and feeding. Emanuel's sister Marta started a summer camp in Finland a few miles from the Nobel summer home in Kirjola. Each year forty to fifty children of the workers spent six weeks in the country. Marta received a medical degree in 1909 from the Petersburg Women's Medical Institute. Surgery was her speciality and she worked first at the city's Obuchovska Hospital and then as a member of the institute's faculty. In 1912 she financed the construction of a fifty-bed surgical clinic persuading her brothers to make contributions—Emanuel donated the iron fence, a younger brother the elevator. When the building was finished she presented it to the institute.

A few years earlier Emanuel had built a much larger structure, a workers' clubhouse on the other side of Lyesny Prospekt where there was a large garden for the workmen and their children with tennis courts and football fields in the summer; in the winter there was a large skating rink and a hill for sledding. The clubhouse was a popular gathering place for workers throughout the Vyborg section, a locale for instruction and entertainment and the site where Emanuel toasted his chiefs and greeted his workers on New Year's Day. It was also where he hosted special banquets honoring particular events or visitors to Petersburg, such as the officers and midshipmen from the Swedish cadet corvette *Freja*.

Minister Gyldenstolpe and the prominent members of the Swedish colony were present along with ranking officers of the Russian navy. A lavish buffet was spread in the flag-and-flower-bedecked reception room with roast bear and wild boar at either end of a groaning table dominated by a huge sturgeon. Blocks of ice overflowed with caviar and other delicacies prized by the palates of two countries. Vodka flowed as freely as the waters of the Neva and the guests of honor shared enthusiastically in Emanuel's generous hospitality. But this was only the preliminary, what the Russians call "zakouska" and the Swedes "smörgåsbord." A fanfare by the Russian military orchestra opened the large double doors to reveal the dining room with formal settings and footmen. And before each plate a bottle of champagne. After the multicourse meal—catered by the res-taurant Ernest—and the usual exchange of toasts the party adjourned to the park for coffee and cognac and a bear cub was presented to the midshipmen as a living memento of the evening. More champagne, this time with strawberries, and many of the cadets were beginning to show the effects of the celebration. No one objected to their indulgence in an old northern pastime—tossing cadets into the air—but when they started heaving Russian admirals skyward and let one drop to the ground their commanding officers stepped in and ordered the drunken offenders back to the ship. Later they were formally reprimanded as were those cadets who had been placed in charge of the bear cub. They lost him while barhopping en route back to the ship. The Petersburg police eventually located the animal and Emanuel expedited his shipment to the Stockholm zoo at Skansen.

The workers' clubhouse was the site of more sober activities—the place where new Nobel machinery was demonstrated along with new engineer-ing and manufacturing techniques. It was where Emanuel and his staff in-troduced the Nobel diesel engine to the Petersburg Polytechnical Society. Immanuel and Ludwig would have been fascinated. So would another missing member of the family, Emanuel's younger brother Carl.

After Ludwig's death Carl had taken over the factory. He was well prepared for the position: an apprenticeship in the Motala yards, two years in a Stockholm machine shop, studies at the Stockholm Technical College interrupted at Ludwig's insistence by a year's work in Baku. In Stockholm Carl was exposed to the talents of another Swedish inventor

Carl De Laval, and he negotiated for the Russian franchise to manufacture and sell De Laval's newest invention, the cream separator. Initially these were fabricated in the Petersburg plant but after 1897, when Russian tariffs on agricultural machinery imports were lowered, the assembled units were brought in directly from Sweden. Six years later a separate corporation was organized to handle increased sales and in 1908 the Alfa Nobel Corporation was formed with Emanuel, a brother Emil, and the Swedish Separator Company as joint owners. It was the sole marketing outlet for cream separators in Russia.

Like De Laval, Carl was also interested in steam engines and, with a college classmate Sixtus Petterson, built a triple-expansion engine to generate electricity for the factory. In 1882 Ludwig had installed the new Edison incandescent lamps but only in his home; Carl's installation powered an electrical system throughout the Sampsonievsky complex. That completed he turned his attention to the internal-combustion engine, working with the oil company engineers to produce a motor which burned fuel directly in the cylinder. Several European factories were working along the same lines. The Swiss Locomotive Factory in Winterthur seemed to have the inside track and in 1892 Carl purchased the rights to manufacture their kerosene-burning engine, personally supervising the fabrication of vertical 3.5- and 7-horsepower engines along with a horizontal one of 10 horsepower. The Russian army was interested and in the next twenty years ordered a hundred units complete with carriages for the cavalry.

Kerosene was an expensive fuel and the Nobel technicians, not really satisfied with their engine, started other experiments and were searching for improved models when Carl died. He suffered a fatal diabetic coma in December 1893 while staying in Zurich at the Bellevue Hotel. A great loss for family and factory. Emanuel referred to him as a "most noble soul and honest person," a man of many interests with "cheerful, kindly temperament."[2] He never was replaced.

Emanuel took over nominal control of the factory as he and Carl's heirs were the sole owners. No other brother was old enough to even consider the job. But he did not have sufficient time and never was vitally interested in factory affairs except as they pertained directly to developments in the oil industry. Yet as his brothers matured and came into the company he staunchly refused to let them step in and prove their own ability to run the plant or even to handle small individual sections of the design or production lines. Having bought out Carl's heirs the decisions were his alone to make, but his caution and conservatism, his resistance to change, were not easy for the family to understand. His brother Ludwig started work in the factory in 1896 and had many suggestions for improvement but Emanuel was not impressed; their sister Marta later complained that he failed to provide the guidance and instruction the brothers so desperately desired. Emanuel, she claimed, had forgotten that he did not have the marshall's baton in his knapsack when he first had started working for Nobel Brothers.[3]

Ludwig managed to force through acceptance of his idea to build a new iron foundry using three- and five-ton Siemens electric smelting furnaces. Even Emanuel had to recognize that this was a valid improvement for it made the factory independent of outside sources to obtain much of the steel later used in assembling diesel engines. Since Ludwig's other proposals were ignored or rejected he gradually lost interest and spent his energies on such vicarious projects as the design of the *Intermezzo,* an eight-cylinder Derens-diesel aluminum racing boat he hoped one day would break the speed record.

His brother Rolf followed Ludwig into the plant ten years later but there was no mellowing of Emanuel's obstinacy. Yet when both brothers suggested that they find employment elsewhere the guardian of the flock was indignant. It might have been the threat to leave, the thought of the embarrassment within the business community, or perhaps recognition of his own stubbornness that eventually led the eldest Nobel to accept another suggestion: Rolf's proposal to install drop-forge presses for the stamping of wheel axles and engine parts.

When Emil's turn came in 1911 he fared better. He was given responsibility for the Alfa Nobel Company selling De Laval separators. Emanuel may have learned from his previous encounters with his brothers or he may have been more impressed with Emil's qualifications: his Stockholm education was followed by work at Motala, by two years with Bessler and Waechter in London, then by a year in Paris primarily to learn French. Emil seemed to adjust more easily to his position in Petersburg, but he was not the man to pick up Carl Nobel's mantle.

At the time of Carl's death in 1893 construction of an economical oil-burning engine was among the most important priorities of the oil industry. One of Emanuel's chief engineers Anton Carlsund, a classmate of Carl in Stockholm hired in 1891 to work on the kerosene engine, reported on a lecture he had heard by an Augsburg engineer named Rudolf Diesel.

He was sent back to Germany to learn as many additional details as possible. Six months later Emanuel was sitting with Diesel in Berlin's Hotel Bristol negotiating exclusive Russian rights for the German's revolutionary engine.

That February Sunday in 1898 was a busy one for Diesel: breakfast with Friedrich Alfred Krupp, a morning meeting with Swedish banker Marcus Wallenberg, an afternoon session with Nobel whom he found to be a "considerate and noble-minded man of imposing, important appearance, with face framed by a short, gray beard."[4] The next day banker and oil baron concluded a contract which led to the organization of Sweden's Diesel Motor Corporation.

But when it came to agreeing on the Russian rights Emanuel needed more time. Diesel pressed but finally agreed to a twenty-four-hour delay. He knew he was dealing with a cautious soul, what he called "an Ibsen man," one whose thoughts and feelings were greater than words revealed. And he knew that "the cold Swede is already more enthusiastic about my motor than I am."[5] For Emanuel it was a question of making the best possible deal for a license to manufacture while retaining the most cordial relations in order to obtain the sole rights to import German-built engines. He knew that diesel motors would increase the demand for oil—his oil—and when the designer hinted that he might contact Rockefeller Emanuel made his decision. Diesel had already concluded an agreement with another American, another baron—this one in beer, Adolphus Busch—so the threat was more bluff than reality, but it forced the ever-cautious Swede to make up his mind. He paid Diesel 800,000 marks for the Russian rights: 600,000 in cash and 200,000 in shares of a new company, the Russian Diesel Motor Company, to be organized with corporate and tax base in Nürnberg. Russian Diesel would grant production rights to the Russian factory.

Diesel was satisfied—he was convinced that he had the better of the deal—and Emanuel was equally certain he had made the proper decision. As soon as he returned to Petersburg he assigned Carlsund and a team of engineers to the project. He maintained close relations with Diesel; a brother-in-law was employed as an engineer in the factory and in the spring of 1910 Diesel and his wife were Emanuel's guests in Petersburg for several weeks. After Diesel's death three years later and on the collapse of his troubled company Emanuel sent money to his son to allow him to complete his education.[6]

Carlsund and the chief designer for the Swedish Diesel Company both believed they could improve Diesel's engine and each developed original adaptations. Hesselman of Swedish Diesel came up with one about a third as large as the prototype and with a capacity in its three cylinders of 120 horsepower. Carlsund's model was similarly reduced in size and his A-type engine was soon being fabricated in two-cylinder 20-horsepower units for the Baku fields and Russian artillery and 60-horsepower units for the factory's own power plant. A lighter and smaller B-type, also two-cylinder, ranged from 75 to 300 horsepower while three- and four-cylinder models produced 600 to 800 horsepower. Carlsund and his staff proved to be an efficient engineering unit, repeatedly demonstrating a skill and flexibility that gave the Nobel factory a commanding lead over diesel developments in other countries.

One of their first installations was in a pumping station on the Transcaucasian oil pipeline at a point near the first steep incline on the eastern section. The government had insisted that all material be of Russian make and Nobel had little trouble winning the contract, convincing the authorities to experiment with a new and less-expensive engine. Nobel also supplied the pumps, Riedler's Rapid Pumps which Emanuel had the Russian rights to manufacture and distribute. Three motors, each of two cylinders and 100 horsepower, were installed and when they proved successful the government ordered eleven additional units for the other pumping stations. Nobel had time to complete only four; the remainder were farmed out to the Kolomna Company.

Other assignments kept Nobel engineers busy. They installed a compressor station in the new Tentelev chemical company in Petersburg. Tentelev's method of producing sulphuric acid had been adopted by Nobel's new Baku plant and for the factory in the capital three 75-horsepower single-cylinder motors were used. Diesels were put also into Petersburg's Perun acetylene factory, into rolling mills in Moscow, textile plants in Tsaritsyn, a cement factory in Gluchosersk. A 400-horsepower engine was installed in the Perm cannon factory, 150- and 400-horsepower units in the central power stations of Astrakhan, Cherson, Korno, Koslov, Livadia, and Moscow, and smaller engines were extremely popular for use in the many flour mills scattered around the Russian countryside.[7] When the Nobel employees arrived for work in the morning they frequently found quite a few kulaks standing on the stairs waiting to place their orders for milling machines. They always paid in advance—the Nobel organization was wary about granting credits for kerosene or engines —carrying their gold rubles in small purses at their side. The Nobel factory was producing more diesel engines than any other concern in the world; for the company which had contributed so much to marine transport it was only natural that it would not ignore the potential of diesel power on water.

It was that veteran of the Volga and sometime visionary Karl Vasilievich Hagelin who made the important breakthrough. He regarded the new engine as an ideal replacement for conventional steam units, the perfect power plant for a series of smaller and lighter river barges he was then considering to use on an eighteen-hundred-mile all-water trip from Tsaritsyn to Petersburg, a route that would eliminate the need for expensive rail transshipment at Rybinsk. It was not a new idea. During the reign of Peter the Great canals had been excavated and numerous small craft were able to make the trip across the lakes and rivers. Emperor Paul I enlarged the canals and improved the lock system at the beginning of the nineteenth century, naming the new canal the Marinsky after the empress Marya Fyodorovna. There were other improvements in the next half-century and another major modernization in 1892–1896. By that time loads of nearly seven hundred tons could be shipped but no one had yet tried to ship petroleum.

Hagelin was certain that a properly designed vessel could navigate the twists and turns of the waterway, risking the frequently stormy Vita, Onega, and Ladoga lakes. In 1902 Karl Vasilievich dramatically proved his point. He unexpectedly appeared in Petersburg one summer day with Volga tug *Votyak* docked on the river just across from the factory.[8] When the excitement died down he announced his intention of building similar ships powered by what he had seen in Sweden and what Carlsund was building in Petersburg, the diesel engine. The capital and the Finnish depot across Lake Ladoga could then be less expensively supplied by diesel-driven barges.

Hagelin contracted with a clever young engineer from the Swedish ASEA company—he later persuaded him to join Nobel—to work on the problem of reversing and regulating the speed which Hagelin believed could be done most effectively by electrical control of the connections between engine and propeller. ASEA generators were coupled to three propeller motors of 110 horsepower and 300 RPMs. He knew he would need that much power to maneuver the twisting, turning Russian waterways. When the units tested successfully, full forward to full reverse in twelve seconds with regulation of speed from 30 to 300 RPMs, he was ready to start the search for a designer.

Johny Johnson of Göteborg was selected and together with Hagelin came up with a light and solid hull, one strong enough to withstand the rocky bottoms along canal embankments. The engine was placed amidships and the electric motors in the stern coupled directly to the propeller axle. On the bridge the captain maneuvered the ship with a power control similar to that on streetcars. For the hold Hagelin designed longitudinal bulkheads running the length of the ship instead of the usual transverse supports. The idea later became standard design on all ocean-going tankers. Construction of the ship was done in Russia. The three 120-horsepower engines were built by Swedish Diesel and ASEA provided the electrical equipment, assembling it under the watchful eyes of two young Russian machinists whose on-the-job exposure in Stockholm made them the best diesel mechanics in the fleet.[9] Once again Hagelin proved the truth of one of his favorite Russian proverbs: "Risk is a noble action."

When the Volga opened for traffic in the spring of 1903 the mother of Russia's waters had a new ship and Nobel Brothers another first. The diesel tankbarge *Vandal* was launched. It was loaded with seven hundred fifty tons of kerosene and navigated the same route to the capital that Hagelin had pioneered the previous year. With a draft of only eight feet, length of two hundred forty-four and a half feet, and width of thirty-one and three-quarters, the *Vandal* easily navigated river, canals, and lakes. Emanuel stood to watch the newest ship of the line glide by the old Swedish fort of Nöteborg, where the Neva flows from Lake Ladoga across a strip of land on which a Swedish king once had erected his line of defense against the Russians—from *Zoroaster* to the *Ludwig Nobel, Petrolea, Anna,* and now the *Vandal*—a proud addition to Nobel maritime history and a proud moment for Hagelin.

The original features designed into the ship surprised the inspector from Lloyd's of London and his sentiments were echoed in the following months by the many technical journals describing in detail these innovations and praising the imaginative creation of Swedish-Russian engineering and shipbuilding.[10]

The *Sarmat* was launched in Tsaritsyn the following summer. Its pair of four-cylinder 180-horsepower engines were made in the Petersburg factory and could be coupled directly to the propeller axle whenever the electrical units were not needed for maneuvering, resulting in a 20-percent increase in power and a considerable savings in fuel consumption. Like the *Vandal* such savings were significant, for Hagelin's new diesels burned 10 to 20 percent less fuel than the conventional steam engines.

By 1906 both Carlsund and Hesselman were working on reversible engines independent of electrical control and the Nobel factory was putting together a three-cylinder four-stroke model with 120 horsepower and 400 RPMs. The success of the *Vandal* and *Sarmat* flooded the factory with orders and additional engineers and draftsmen had to be hired. The payroll grew to more than a thousand and new buildings were erected. Emanuel's younger brothers Rolf and Ludwig, now part of the management, had their own ideas about modernization of the administrative and technical routines but Emanuel resisted them and refused to entertain any proposals for change. He also opposed Hagelin who wanted to follow up his *Vandal* and *Sarmat* successes with diesel conversions of larger ships.

Karl Vasilievich had had ample time to consider the possibilities and was tearing at the bit, eager to put his plans into action. After the events of 1905, his nerves frayed and body exhausted by the strain and horrors of Baku, he had taken Emanuel's advice and traveled for several months in the United States accompanied by a Nobel geologist and touring American oil installations. On his return he settled down for a time on a large estate south of Stockholm—he had dreams of becoming a country gentleman. But a few months of ease and relative inactivity convinced him that he certainly was not ready for retirement and his mind turned toward resuming his Russian career. When the Swedish government asked him to be their consul general in Petersburg he needed no more prodding, especially after they promised to provide him with a staff of three and full freedom to return to his director's responsibilities with Nobel Brothers.

He was again working with Johny Johnson in 1907, this time drawing up plans for a forty-five-hundred-ton diesel tanker with two 500-horsepower engines, four times as large as anything then on the Caspian. Emanuel and the board rejected the idea. Hagelin would not be denied and went to another firm, Markulyev Brothers, which had recently signed a five-year contract to sell and distribute Nobel products. With that contract in hand, their good reputation, and Hagelin's presence and persistence, the Kolomna shipyards agreed to advance credit for construction, believing with Hagelin and the Markulyevs that the new ship would soon pay for itself just as Ludwig's *Zoroaster* had done so many years earlier. Johnson worked with the Kolomna engineers who were under the skilled supervision of a Pole named Korejvo, a man who agreed with Hagelin's suggestion to build first a paddle-wheel diesel tug and then the tanker, using the tug as model for experimentation to solve the problems of nonelectrical control of the large engines. By the summer of 1908 they had the solution. Sliding down the slips of Kolomna came the world's first diesel tug and a short time later the world's first diesel-powered tanker. With a touch of poetic self-justification Hagelin christened them *Mysl* (thought) and *Dyelo* (action).

There was no doubt about the value or success of Hagelin's stubborn scheme. He was right and the board was wrong. Emanuel immediately put Karl Vasilievich in charge of a sweeping program to modernize the fleet. Twelve million rubles were budgeted and when Hagelin used up that amount additional millions were forthcoming. By the outbreak of war the job was almost finished: new ships, tugs, barges, many built at Kolomna where Sokolov, Hagelin's boyhood friend from Saratov and former chief of the Rybinsk depot and transshipment point, served as Nobel project officer and general comptroller.

Of course there were problems, some of them major like the improper construction of cylinders and gears, the difficulty of finding a reliable method to reverse the engines. But Hagelin, Carlsund, Johnson, and the Kolomna engineers hammered out solutions, found shortcuts, and drew up blueprints for new diesels as soon as the first models were off the line. Some ships like the *Robert Nobel* were converted and five new tankers were launched including the first diesel on the Caspian, the five-thousand-ton *Emanuel Nobel*. There was also a second *Zoroaster* considerably larger than the world's first tanker and powered by a pair of 500-horsepower engines. Several smaller ships were built for the Ob River—Nobel oil was moving in bulk through Siberia—and, appropriately, a five-thousand-ton *K. W. Hagelin*. Of the sixteen motorships in the world powered by engines of 600 or more horsepower in 1912 fourteen were Russian, outfitted with Nobel diesels. By 1915 Nobel captains were in command of three hundred fifteen vessels displacing a total of three hundred sixty-eight thousand tons. Only the Russian navy had greater tonnage in the empire and they too were modernizing their fleet.

Displaying a farsightedness unusual for the tsar's military planners the navy, forced to rebuild and replace the losses suffered during the war with Japan, commenced a program of conversion to diesel power. The Germans also were putting diesels in some new submarines and the British and Americans were making noises about modernizing their own fleets and utilizing oil, but they were far behind the Russians who ordered their first diesel ship in 1905. The ministry's initial contract with the Ludwig Nobel factory called for thirty-two engines to equip the gunboats of the Siberian Amur fleet, decimated by the Japanese in that "glorious, little war." But the plant could not meet the navy's delivery deadlines so the order was shared with Kolomna using Nobel designs and patterns. The first of the gunboats, the *Skval,* was launched in the spring of 1908 and additional orders were placed when it was tested to the navy's satisfaction. Among the many Nobel-equipped ships built in the next few years were the Caspian gunboats *Kars* and *Ardagan,* the Black Sea cruiser *Yastreb,* the icebreaker *Galerny.* Nobel also manufactured electric generators for naval power stations, auxiliary engines for cruisers of the *Navarin* class, and the power plant for the first diesel passenger-ship on the Caspian, the *Empress Alexandra.*

In contracting with the Nobel factory for the diesel-powered gunboats the navy also ordered two additional units for a submarine being built by the Baltic Company. Two years earlier the ministry had commenced a ten-year plan of submarine construction by placing orders with Baltic and the Neva Factory as well as with Krupp and an American firm, but Nobel's pair of 120-horsepower engines were the first diesels ever used in a submarine. There was a delay in design and fabrication and the first unit was not ready until July 1908, the second three months later. On October 23, 1908, the world's first diesel submarine, the *Minoga,* was launched. One hundred seven feet long with two front torpedo tubes and a cruising speed of eleven knots, the submarine served as prototype for the vessels which followed.[11] The following year came the three-hundred-seventy-ton *Akula* armed with four torpedo tubes and driven by three 300-horsepower Nobel engines; it was selected as the most reliable and seaworthy underwater vessel in the fleet and became the model for all later submarine construction.

And there was considerable construction. The ten-year program was revised and expanded and a new company chartered to handle additional orders—the creation of M. S. Plotnikov, a member of the board of the Discount Loan Bank and managing director of the Lessner Company. The new firm was built near the naval base in Reval (now Tallinn) where the waters were deep enough for diving and torpedo tests. Plotnikov felt that the Lessner Company's experience in manufacturing mines and other naval armaments was sufficiently impressive when combined with Emanuel's engines and expertise and bank credits of 5 to 6 million rubles to allow a new organization to compete successfully with the traditional suppliers and builders, the Baltic and Neva companies. He was certainly confident of his influence within the ministry. A postwar naval commission reported that Plotnikov "definitely knew everything that was said and done in the ministry about the problems which interested him" and was shrewd enough to be able to evaluate correctly the impact of the opinions and decisions of various officials. And there was no doubt, the commission claimed, "that he could not only predict the results but by timely bribing could also guarantee a decision in his favor."[12]

By timely dangling of higher salaries he could woo such experts from the Baltic factory as Major General I. G. Bubnov, professor at the Naval Academy and the country's outstanding submarine designer. His brother, an engineer at Baltic, and thirty-six other experienced hands joined him in September 1912 as the ministry decided to award the new company—entitled Noblessner—the contract for eight of the twelve new submarines. But there was still no actual company. Only when the signed contracts were in hand—they were worth nearly 15 million rubles—did Plotnikov and Emanuel charter their new enterprise. Noblessner shareholders included the chief of the ministry's administrative and technical section who owned six hundred shares, and their first meeting was held four months after receiving a confirmed government order.

Construction at Noblessner's Reval yards did not get under way until the spring of 1914 and by the time war started the Russians had a total of twenty-four submarines under construction and twenty-two deployed in the Baltic, Black, and White seas. Swamped by a multitude of military orders the Nobel plant was unable to supply enough engines for the new submarines, even for those on the slips at Reval. Baltic's order for German engines was obviously cancelled by the war and those received from the English and Americans did not prove satisfactory. The navy was forced to use the Nobel engines installed in the Amur river gunboats while ordering the Nobel factory to expedite construction of larger more-powerful units for larger more-powerful submarines. The result was the design of the plant's two-stroke diesel engine but only two submarines were ever equipped with it, the *Zmeya* and *Kuguar,* both assigned to the Baltic fleet. The Kharkov Steam Locomotive Company on license from Nobel built and installed similar engines in submarines destined for the Black Sea, and in Riga the Felzer factory started building Nobel diesels but the advancing German army forced evacuation of the plant to Nizhni-Novgorod.

Evacuation was practically the only problem the Nobel factory did not face in the hectic war years as the payroll exceeded two thousand, new buildings were constructed, and one urgent government contract after another was issued to the company. The Petersburg plant was working on so many military orders it had little time for modernization of its own fleet.

Hagelin had to rely on Kolmna and Swedish yards. When the company decided to convert its Volga steamer *Yakut*, originally built by Motala, two engines from the Swedish Diesel Company were used. When Hagelin wanted to build an ocean-going tanker it was Hesselman who did the designing. The double six-cylinder engines with a combined horsepower of 3,000 were the largest Hesselman's company had ever assembled and the ship was the first of a long line of noted diesel transports built by Sweden's Gota Shipyards. The last one constructed for Nobel Brothers was the *Varyag*, an ocean-going tanker which Hagelin sold on the spot as soon as it was launched for 5.3 million Swedish crowns. With that sum in hand he paid Gota the 1.8 million crowns the ship cost to construct. That was in 1916, with wartime inflation and speculation and with severe shortages in marine transport.

The Nobel factory was on round-the-clock shifts which left little time for research and development. But Carlsund was able to perfect his two-stroke engine, an imaginately designed unit that developed twice the horsepower of the four-stroke model. One drove the factory lathes used in turning grenade casings. The brilliant Oscar Derens also found enough peace to complete the design and fabrication of four- and six-cylinder diesels using one of his units, a 200-horsepower engine, in the miniature submarine *Delfin*. Smaller motors were put into smaller submarines designed for harbor defense. Derens was occupied primarily with the blueprints and construction of Emanuel's new yacht, a sleek hundred-foot motor cruiser built on the lines of a small destroyer and powered by a pair of 450-horsepower diesels capable of twenty-six knots. This was the *Gryadoustiy*, the Future One, and Emanuel had it built to impress the navy with the latest Nobel diesels.

Emanuel took a direct personal interest in the design and construction of his yacht which was also intended as speedy transport for family and friends from the banks of the Bolshaya Nevska, across from home and factory, to the country estate in Kirjola, a journey of just six hours on the *Gryadoustiy*. But before both engines could be installed the vessel had to be taken for safekeeping to Vyborg away from the clutches of the rebellious sailors at Kronstadt.

The achievements of his engineers and assembly lines in supplying the navy were a source of pride to Emanuel, but as the factory grew it became much more difficult to control. Already overtaxed by the mounting responsibilities of the oil company he had less and less time to devote to the plant. On the fiftieth anniversary of the factory in 1912 he decided to

convert it to a corporation with himself and his three brothers as sole
shareholders. Emanuel reserved the power and sole authority for major
decisions. Nothing really changed. He remained conservative in his
judgments fearing an overextension of assets, worrying about the sudden
expansion of plant and personnel, perhaps remembering what had
happened to grandfather Immanuel when he mortgaged his financial
future to the promise of government contracts. Over the course of several
years he had to make up plant deficits totalling 2 million rubles and he
found it increasingly impossible to take the time from oil-company
business to oversee factory activities. Ludwig Nobel, Incorporated, needed
tougher management and it would continue to need massive transfusions
of capital. He therefore decided to offer half of the share capital to the
Petersburg Discount Bank. One of their team's first actions was to float a
2-million-ruble loan for Emanuel's reimbursement.

By then the nation was at war and the factory was at the disposal of the
Ministry of War producing, in addition to the diesel engines for ships and
generators, a variety of military supplies: picks and shovels for the engineers, wheel rings for trucks and caissons, fuses for 75-millimeter shells,
grenades, and artillery ammunition. By that time Emanuel was totally involved with the effect of the war on the fortunes and future of Nobel
Brothers.

CHAPTER FOURTEEN

One Step Forward, Two Steps Back

The decade between revolutions, between the Grand Rehearsal and the holocaust which would bring total military defeat, economic and social collapse, terror and civil war, was for the Russian oil industry a time of crisis, consolidation, and confusion. The title of Lenin's polemic—"One Step Foward, Two Steps Backward"—described perfectly the situation. Recovery from the slaughter and destruction of the dark days of 1903–1906 was distressingly slow. Beset with a multitude of postwar and postrevolutionary political and economic problems it was unwilling at first to confront, the government knew that much of the nation's recovery depended on a healthy petroleum industry delivering fuel to the factories again starting up across the land. But that industry, in the words of Minister of Finance Kokovzev, was on "the brink of disaster." Still the government was painfully slow to react. It did not expedite completion of the final phase of the Baku-Batum pipeline or lower the railroad freight rates which finally had been stabilized in 1900 using a figure based on the previous ten-year average, but were sharply increased at the beginning of the Japanese war.

It was left to Emanuel and his lieutenants to show the way. The editor of *Petroleum World* observed that the Nobel Company was "the popular leader in all movements for the improvement, and in the present crisis, for the reconstruction of the industry at Baku."[1] But the government had to do something. In addition to fueling the factories a revived oil production was needed for trade—only wheat earned more foreign currency in the marketplaces of the world. A meeting of the oil industrialists at last was called in Petersburg in October 1905.

Kokovzev opened the conference with an appeal for the cooperation of all producers large and small. He said nothing about the causes of the trouble or what the government proposed as remedy or how they would work to prevent possible recurrences. To Emanuel those factors were obvious: widespread worker discontent at the lack of progress in the Baku reform program which was supposed to provide better housing and working conditions. But he did not mention in his statements to the conference the other issues: the mutual hatred of Armenian and Tatar, the role of the Black Hundreds, the inefficiency of tsarist administration. Those were the causes cited by his opponents—in many cases the very industrialists guilty of the worst abuses of the workers.

The Oil Producers Society placed the blame squarely on the government and demanded immediate restoration of order, military protection of property, and statutory regulation of relations between capital and labor. It also requested interest-free loans to enable the smaller firms to rebuild. The government eventually made available 20 million rubles but charged 5 percent interest on the money. The society was powerless to force the issue; the government was in a position to counter any general offensive by the producers with tactics which would split the group into its three constituent and basically contradictory camps—the producers, the refiners, and those large companies engaged in both activities. Agreements on administrative and social affairs—record-keeping, subsidization of fire departments and hospitals—could be forged by society members, but when more substantive matters were on the agenda lack of unity was immediately apparent and individual interests withdrew into the shadowy security of the time-honored Baku traditions inspired solely by jealousy and greed.

After the Grand Rehearsal the rules were changed. The government did not want to split the industry but to unify it, to force cooperation of all segments in order to gain maximum strength for reconstruction. A new Council of Thirty was selected jointly by government and industry to replace the old executive committee as spokesman and negotiator. But the change did not achieve the desired improvement; instead of the Big Five there were now thirty, and whenever one or several of that number did not agree with a given position or vote they merely withdrew from the council and refused to honor the decision. Without enforcement or penalty provisions the industry was impotent, unable to create from the Council of Thirty a common front, to marshal the power of a key sector of the economy. The inevitable result was continued postponement of urgent reform in field and factory, refusal to come to grips with the basic problems: reconstruction of damaged properties, cessation of the insanely uneconomic exploitation of oil reserves by the elbow-to-elbow drilling on tiny plots of land.

The more reactionary elements in court and country regarded this inability to cooperate as an inevitable result of the evils of free enterprise, an integral fault of the capitalist society, a cancerous import from abroad spread across Holy Russia by foreigners and Jews. To many of the orthodox the foreigner was often equated with the Jew, deserving of similar distrust and disrespect. Wherever the orthodox cast his suspicious glance there were foreigners and Jews: Englishmen in the iron mills, Belgians in steel, Germans and more English in textiles, French in copper, Swedes in petroleum, and everywhere the Jews with their banks supporting all those foreigners—the same Jews who had financed the enemies of the empire in the Crimean War and fifty years later the enemies of the empire in the war with Japan.

The conclusion was obvious. Cleanse the country of the capitalists, expel the Jews, replace anarchy with autocracy. Nationalize the foreigners' holdings. Raise the protective hand of the tsar over the economy. Take over those industries profiting on the exploitation of Russia's natural resources. Convert the oil industry to a state monopoly.

Not the cries of Communists but the demands of the most conservative forces in the land. In 1906 the Duma dispatched a delegate to Baku empowered to survey the area and report on whether or not the industry should be nationalized. His positive recommendation was accepted and approved by the Duma, then set aside by the ministry.[2] By that time the immediate crisis had subsided, the fears of resurgent revolution calmed, and Stolypin's economic reforms—a pragmatic fusion of Witte's ideas and traditional peasant economy—were having dramatic effect on the recovery of the country. There was a period of general prosperity. The annual rate of growth in industrial output exceeded that in the United States, Britain, and Germany. Millions of peasants became individual landowners, there was a greater freedom of movement within the country and, for the three million workers, some meaningful progress in education and welfare including health and accident insurance. The revolutionary spirit waned.

Stolypin's economic revitalization was being fueled by Baku oil as well as by Donets Basin coal. The new Russian black gold from fields opened the year after the revolution was not of the highest quality nor was it as efficient as fuel oil, but with oil in short supply and its price driven to new record levels Russian industrialists were grateful for the alternative.

Recovery in Baku was delayed, high prices remained, and Witte's policy of appending the oil industry to the Russian body economic was exposed as a cancer to the transport and manufacturing industries. In 1907 the average price of fuel oil rose to 29 kopecks a pood—hitting 43 kopecks during one period, four times the 1903 cost. At its cheapest levels after the 1905 revolution it was still twice the price of Donets coal. The manufacturers in the north converted their equipment to coal vowing never again to be caught in what they regarded as an unjustified price squeeze by the Baku producers.

The larger oil companies were the greatest beneficiaries of underproduction. For the industry as a group this meant in 1906 a profit of 115 million rubles on 7.25 million tons. Two years earlier the profit had been 90 million rubles on 9.7 million tons. The Nobel dividend of 10 percent in 1904 was raised to 12 percent the following year, 18 percent in 1906, and 20 percent in 1907. Nobel then had nearly a third of the 1.4-million-ton capacity of Baku's leading twenty refineries—the smallest of the twenty could refine as much oil as Nobel did in 1889.

Windfall profits and lowered production meant more than dividend increases; they meant a new era for the Russian petroleum industry. The peak years—Russia's position as leading producer in the world—were now history as Baku oil no longer was able to capture world markets on price and volume alone. The age of the Baku fountain also was over. What the Black Hundreds and the rampaging Tatars could not achieve was done by the forces of nature. Gas pressure on the Apscheron peninsula had dropped to the point where more expensive techniques were required to bring out the oil: bailing and deeper drilling. Before 1900 the deepest well had been seventeen hundred fifty feet. By 1909 it was twenty-five hundred. The average of the Balakhany-Sabunchy wells was fifteen hundred eighty; twenty years earlier it had been six hundred sixty-five, and in the first year of Nobel Brothers only two hundred seventy-three feet.[3]

By 1911 the American system of rotary drilling was spreading across the peninsula as the most effective way to reach the deeper strata. Nobel had nine rotary rigs in service and one of their subsidiaries three more; the competition had a total of twelve. The new system was more expensive but increased drilling costs were added to the ultimate cost to the consumer. As that price rose buyer resistance and reliance on coal spread and there was a period of unemployment—by 1910 more than a quarter of the workers were laid off. But the depression was short and after full recovery and adjustment in the marketplace the workers felt strong enough to stage a strike. The result was a 20 to 25 percent increase in wages—the workers were sharing in some of those inflated profits.

Nobel was spared the worst effects of the depression: the company was so large it could absorb such shocks, shifting labor to other projects. By their thirtieth anniversary in 1909 Nobel Brothers had thirteen different factories covering two hundred fifty acres in Black Town alone—five refineries, five installations producing chemicals, a gas works, a sawmill, a barrel factory, and a mechanical workshop. The refineries were reputedly the finest in the world and consisted of seven separate buildings for distillation with giant steam preheaters, fifty-one boilers and coolers, and thirty-one storage tanks. In the cleaning section were twenty-six other tanks, a half-dozen mixers, seventeen pumps, a great variety of motors. The two-ton refinery of Robert Nobel would not have taken more than a small corner in any one of these buildings.

The smaller producer, the little refinery, could no longer compete and that meant still less production. The results on the export trade were disastrous. The year after the revolution, when domestic demand exceeded the supply, Russian petroleum exports were only one-sixth as great as the American. In 1903 they had been nearly two-thirds. Annual exports from Batum which had been averaging 1.3 million tons dropped to less than a half-million. The troubles in Russia, the riots in Baku, the gradual depletion of the known Baku fields—the Second Thirty Years War was entering its final phase with the Americans commanding the heights and launching concerted attacks to gain back territories lost to Nobel, Rothschild, and Mantashev. For the Russians surrender was the only option. They could hope only to come out of the final negotiations with a viable marketing organization still intact, looking forward to the time when political and economic conditions at home once again allowed them to take the offensive abroad.

Hans Olsen attended a January 1906 meeting in Frankfurt called by the Deutsche Bank with this in his mind. After the 1905 Treaty of Björke the Germans were now Russia's "most favored nation" and in control of some two-thirds of its foreign trade—some Russians felt they had been reduced to the status of a German colony—and the Deutsche Bank was concerned about the eventual consequences on the general European oil market of the chaos in Russia. The bank hoped to forge a closer cooperation between Nobel-Rothschild and their own interests. Olsen represented Nobel; Aron, Baer, and Fred Lane—now a Shell director and head of the London syndicate backing the Schibaieff Company—spoke for Rothschild. The representatives of the Rumanian industry also were there, eager to carve a larger slice of the European market at the expense of the crippled Russians.

It was that issue which divided the meeting. Olsen would not agree to the Rumanian demands and resisted the pressure from the meeting's chairman, President Arthur von Gwinner of the Deutsche Bank which was one of several German institutions heavily committed in all sectors of Rumanian development and which owned the majority of shares in the largest Rumanian petroleum company Steaua-Romana. The Berlin Discount Bank had major interests in other Rumanian oil firms and provided backing for the German Oil Corporation with its own Rumanian holdings. Rumanian oil, like the Russian, supplied the Germans with alternative sources and the means to break Standard's monopoly. The Rumanian and Russian production enabled the Deutsche Bank to organize in 1903 the Deutsche Petroleum Aktiengesellschaft (DPAG) with a capitalization of 35 million marks and the backing of important European banks including the Swiss Credit Bank, the Vienna Bank Association, and the Central German Credit Bank.[4]

The Germans were erecting a Central European economic bloc and Rumania was a key country in that bloc. They had watched with considerable apprehension the 1904 overtures by a combine of Nobel, Rothschild, and Mantashev bidding on Rumanian crown lands, hoping to develop the oil potential as an exclusive all-Russian concession. These bids exceeded those previously submitted by Standard but were still not acceptable. German preeminence was thereby guaranteed and a Franco-Russian economic wedge on their southern flank in the persons of Nobel and Rothschild was deflected.

It was far better to have that wedge in Germany itself, to keep it under the control or at least under the watchful eye of the Deutsche Bank while simultaneously erecting an orderly superstructure to provide the unity and strength to meet the onslaught from the United States. The first meeting in Frankfurt failed to achieve those goals but at a second held two months later in Paris the same representatives spent ten days in Rothschild's Rue Laffitte offices hammering out compromises on the basic issues. The Deutsche Bank legal staff then retired to draw up the documentation. By June they were ready and the delegates reassembled in the imposing Berlin headquarters of the Deutsche Bank. Von Gwinner again chaired the conference and led the signing of papers creating the European

Petroleum Union (EPU). The organizational principles were similar to those of SAIC, Nobel's distribution mechanism for lubricants: market and delivery quotas were established; depots and tanker fleets pooled; ownership divided among the participants with Nobel holding 36.19 percent, Rothschild 29.61 percent, and the Deutsche Bank 28.2 percent. Kerosene was the primary commodity but all petroleum products were covered except lubricating oils.

Fortified by the power of the Deutsche Bank the Russians and Rumanians now were able to deal with the Americans who could no longer take advantage of price-cutting in an individual country, eliminating one or another producer from the market. Standard was confronted by an organization able to move with speed and unified strategy equal to its own and it soon came to terms with EPU. The two years after the Berlin agreement witnessed one settlement after another across the continent as Standard distributors and subsidiaries signed with EPU, ending the price wars of a quarter-century. Each side was satisfied with the arrangement. With the continued shortage of Russian oil for export EPU had no surplus to flood any area and its members were content with their guaranteed 20 to 25 percent of the European market. The executives at 26 Broadway were more than content with 75 percent.

Similar market-sharing agreements with Standard were made in Britain between the American affiliate and the newly organized British Petroleum Company (BP), a merger of Nobel and Rothschild's Consolidated Petroleum Company with General Petroleum, the British outlet of Deutsche Bank and Rumanian interests. BP was the equivalent of EPU; Emanuel and Olsen became directors of BP just as they were of EPU.

During the many meetings which led to the formation of EPU and BP the Rothschild and Nobel relationship, in the view of Olsen, "could scarcely have been friendlier or more confidential."[5] Baron Edmond —Alphonse had died in 1905—told Olsen privately that Nobel Brothers should take over all Rothschild holdings in Russia as well as related properties in Europe, operating them as part of an expanded Nobel Company with Rothschild in the role of shareholder. It was precisely the scheme which Olsen and Emanuel had discussed with Standard but it was not nearly as tempting. Alliance with Rothschild would not add to the trade and sales advantages already enjoyed by Nobel through existing agreements, there would be no great saving in expenses, there would be a significant strain in the already tight financial conditions with added commitments in a time of great political uncertainty. Finally, there was no assurance that the government would welcome a Nobel colossus. Emanuel informed the baron he was not interested.

Olsen had recommended such a rejection but at the same time he advised Emanuel to consider another proposal from Edmond: the entry into Far Eastern markets in league with Rothschild, Shell, and Royal Dutch. The Paris house had been involved in that area ever since their first contract with Shell in 1891. Shell's superior transport and storage network in the east provided the marketing Bnito needed and when Shell and Royal Dutch were negotiating their merger Rothschild was offered an important piece of the new corporation. Rothschild then offered half his share to Nobel.

Olsen was excited about the prospects of joining such a powerful alliance, of gaining a firm foothold in the Orient. But Emanuel was against it. He was not very fond of the Samuel brothers and their Shell associates—they had fought effectively against his entry into the Asiatic Petroleum Company a few years earlier. There also might have been an undercurrent of anti-Semitism in his strong reactions but this had certainly not been revealed in all his dealings with the Rothschilds who were hardly good Swedish Lutherans; Aron's father was a rabbi in Strasbourg. A stronger factor was probably Emanuel's lack of interest in the area and his total ignorance of it. Perhaps he was influenced by the tragedy of the Russo-Japanese War, the various penalties paid for the tsar's bungling Far Eastern policies.[6]

Emanuel's only previous exposure to the area had come through his East Asian Oil Trade and Industrial Company which he organized primarily to service the fuel oil needs of Witte's Transsiberian railroad and the Amur River flotilla. From the opening of the line in 1896 it was apparent to the Petersburg planners that fuel oil would have to be used; without it east of Lake Baikal the engines had to shift to wood, a scarce commodity in those parts, or to brown coal, an unacceptable substitute. Ironically, the Baikal area was later found to contain major oil deposits, but at the turn of the century the only petroleum ever seen that far east came in Nobel tankcars and was stored in Nobel depots. East of Irkutsk in the farthest reaches of the empire, in Transbaikalia, Amur, and the maritime provinces, the Nobel product was peddled to the limited market and pumped into military and railroad reservoirs. But to Emanuel this was not really a Far Eastern market but merely an extension of the conventional Russian sales areas. Export to Manchuria was a similar extension into a Russian sphere of influence; Nobel shared the Manchurian market with Royal Dutch after a local oil war had eliminated Standard which then concentrated on North China.

There was another factor at work to cool Emanuel's interest in an active partnership with Royal Dutch and Shell. That was the personality and the power of Henri Deterding. Lack of alertness in dealing with that Napoleon of oil was not only dangerous; it could be fatal, forcing assumption of a definitely subordinate position—as Marcus Samuel well knew. Nobel had no desire to place himself in that kind of relationship either directly or through an emissary like Olsen. And after Olsen retired in 1908 there was no one else even to consider sending to the heights to negotiate, barter, and argue with Deterding.

Exhausted physically and overwhelmingly pessimistic about the future of Russia, Olsen resigned his post as director and also as Norwegian consul general—a position he had held since 1905 when his country gained its independence from Sweden. He had informed Emanuel of his intention to retire nearly a year earlier but had probably made up his mind before that. After the events of 1903–1906 he had little faith in the future stability of the country and convinced his wife to sell most of her Nobel stock. The million Swedish crowns from the sale were put into Scandinavian banks and government bonds. He returned to Kristiania (now Oslo), built a mansion, organized the Andresen Bank, and served on a number of corporate boards including Norsk Hydro. The Rothschilds were the largest of the French group of investors in that important company and Olsen's experience in dealing with them was of great value. He continued to attend the meetings of EPU and BP.

As his replacement Olsen selected Ernst Grube, a Petersburg native of German parentage with a solid command of languages, a commercial education, experience in the State Bank and—through Witte's influence—some background as government financial expert assigned to Teheran. He did not have the commanding presence of Olsen and perhaps not the great tact and courtliness, but it did not appear at the time that he would be called upon to play the Olsen role of diplomat and international negotiator. With Emanuel firmly refusing to become involved in the Far East, with EPU and BP solidly established and the general export levels still depressed, the main thrust of Grube's efforts had to be domestic. The Second Thirty Years War had ended and the company had to concentrate its energies on developments within its own borders.

The focus was now changed; Russian producers no longer looked to

Europe for creation of new and larger markets to absorb the great sur-
pluses. There were other suppliers to meet the demands of the age of the
internal combustion engine: the new Texas, California, and Mexican
fields, those in the Netherlands East Indies, and the still small play to the
south on the Persian Gulf. The Anglo-Persian Company was organized in
1909 and within three years gulf production reached forty-three thousand
tons; by the outbreak of war it had increased sixfold. But the Europeans
were still looking to Russia with new hopes of profits.

New fields were opening three hundred miles north of Baku in Grozny,
a hundred miles closer to a Black Sea port Novorossiisk; in Chatma near
Tiflis; on the northern shores of the Caspian near the Ural River; and in
Maikop nearly three hundred miles west of Grozny and only fifty miles
from the Black Sea port of Tuapse. It was the English, late off the mark in
Baku and undaunted by the disappointments of the Berekey boom, who
rushed to take advantage of the new opportunities, especially in Maikop.
London investment trusts and joint stock companies proliferated at a
dizzying rate. Their 1905 investment total of nearly 100 million rubles in-
creased sharply as more than sixty new companies were organized in-
cluding the Anglo-Maikop Corporation, the Maikop and Eastern Oil
Company, British Maikop, International Maikop, London and Maikop,
Maikop Midland, Maikop and General, Maikop Central, Maikop
Premier, Maikop Shirvansky, Maikop Spies, Scottish Maikop, Maikop
Standard, and then the Ural Caspian Oil Corporation and the Chatma
Oilfield Company. The *Pall Mall Gazette*, recognizing that "the speculative
fancy has been tickled by the remarkable discoveries on the Maikop
fields," published a pocket guide for the investor listing thirty-seven major
firms and warning the reader that "strongly divergent opinions have been
exposed as to the future of the oil share market," emphasizing that "the
promotions are quite sufficiently plentiful and much discrimination is
needed."[7] The caution was well-advised. By 1916 only five companies
were still in business.

As Maikop production increased the importance of Batum as transshipment terminal declined. Already seriously affected by the drop in Baku
production and the domestic demand funneling most of that oil away from
the export market—less than 10 percent of Baku oil was moving along the
pipeline or the Transcaucasian railway—the Black Sea port faced competition from the northern ports of Tuapse and Novorossiisk. Batum had
not been heavily damaged during the troubles of 1903–1906; only the
Armenian buildings had been destroyed. But just about the time they

were rebuilt and back in operation Bnito decided to shut down its Batum installations. With two-thirds of all Batum's exports consisting of petroleum products this withdrawal of a major company was a serious shock to the local economy. The depression of 1910 was another setback, but even in that year nearly three million cases of kerosene were shipped from the port.

The company which had started the boom back in the 1880s dismantled its tank farm, moving it to Maikop, and sold its case and barrel factories. Several thousand workers were suddenly on the streets but—much to Rothschilds's credit—every one of them received some kind of pension or a half-year's wages. Bnito shifted its interest to Maikop and to Grozny where it participated in a joint venture with Royal Dutch and Shell, the Nouvelle Société du Standard Russe de Grosny, known as Russostand. Capitalized at 12 million rubles the new firm was British-registered and administratively supervised, but when it came to decisionmaking it was all Deterding.

And it was Deterding, ever the expansionist, who negotiated in 1911 with the Rothschilds for the purchase of all their Russian properties. Nobel was not interested in merging with Rothschild, in taking over Bnito, Mazut, and the extensive infrastructure, but Deterding certainly was. Twenty-seven and a half million rubles were paid for Rothschild's 80-percent ownership in Bnito and Mazut, for the fleet of ten tankers, forty barges, the eight hundred thousand tons of storage capacity, an annual production of four hundred thirty-six thousand tons of crude and three hundred thirty-eight thousand tons of refined.[8] It was less than half the size of Nobel but certainly an impressive, valuable addition to the Deterding empire.

For Emanuel the sale was a serious setback. The rather close and mutually advantageous relationship which he and his associates had forged with Rothschild representatives over the years did not carry with it a guarantee that similar rapport could be established with successors. Worse, Emanuel's acknowledged position as senior and most powerful member of the Nobel-Rothschild alliances would certainly be threatened by Deterding, the Napoleon of the oil industry, a recognized world figure incapable of following behind Nobel or any other man. With the ingestion of Rothschild's Russian holdings his Royal Dutch Company commanded some three million tons of oil a year—1.6 million from the Netherlands Indies, a half-million each from Rumania and Baku, and three hundred eighty thousand from the new fields in Grozny. As soon as the Rothschild negotiation was completed he directed his attention to the new Ural field on the Caspian, planning for still more oil.

For Rothschild it was another financial coup. Twenty-seven and a half million rubles of stock in the two companies, 60 percent in Royal Dutch and 40 percent in Shell, the same proportion as their 1907 merger. Not cash which had to remain inside Russia or shares in some newly organized Russian company, but stock in a solid western organization with a very bright future and brilliant leadership. After twenty-seven years of profits in the golden age of the Russian oil industry the Rothschilds knew when to withdraw. In a few more years they would have lost everything.

Deterding's debut on the Russian stage did not pose the greatest threat to Emanuel during this period, however. That came from an independently organized group, the Russian General Oil Corporation (RGO). Chartered in 1912 with a healthy $12 million floated half on the Russian market and half on the British, this English holding company with a prestigious collection of Russian directors in a surprisingly short time gathered under its corporate wings the Russian independents—Mantashev, Lianozov, Mirzoiev, Goukassov—and a valuable infrastructure of tankers, depots, and refineries to handle their massed production. By 1916 the share capital totaled more than 120 million rubles and the total production was nearly that of Nobel and Royal Dutch–Shell combined. RGO was a new and important third force to compete with the two giants and it enjoyed substantial backing from three important Russian banks: the Russian-Asiatic, the Petersburg International, and the Siberian.

It was an echo of Witte's grand strategy of twenty years earlier, a balance of forces within a crucial industry. But RGO had some serious intrinsic weaknesses probably inevitable in such an assemblage of disparate interests, and its ambition soon exceeded its power. Seeing an opportunity to reach for the stars RGO started buying Nobel shares, mainly on the Berlin Exchange, bidding up the price with money advanced by the banks. By 1914 the group felt that its 6-million-rubles-worth of stock was enough to make the move. At the annual meeting of Nobel Brothers its representatives challenged the leadership and tried to vote out the board of directors. But the loyal stockholders and a series of banks stood by Emanuel and his associates.

It was a costly tactical error for RGO. Not only had the corporation paid inflated prices for Nobel shares; it failed to purchase enough of them to execute its plan. Six months later the director of the Petersburg International Bank—the son of former Minister of Finance Vyshnegradsky, a director of the Kolomna shipyards and a good friend of Hagelin—offered Nobel all his RGO shares. He was a member of the RGO board of directors as was the head of the Russian-Asiatic Bank who also offered his RGO shares to Emanuel. The two were members of Nobel's board and their experience with the two companies convinced them Nobel had the greater potential and should be supported as leader of the Russian oil industry.

Emanuel was slow to react and when he did make up his mind he told his executives that no matter what they did to persuade him he would never agree to buy shares in RGO. Hagelin and the other directors worked on him and Hans Olsen came to Petersburg to study the situation, but not even his firm recommendation could force Emanuel's hand. It was another year before he agreed to the purchase and paid out 20 million rubles for the banks' shares. Emanuel's decision put Nobel Brothers in command of more than half of the Russian oil industry.

In addition to the investment in RGO—not a majority ownership but adequate to the purpose of monitoring developments and preventing future raids—Nobel through other investments controlled or held leadership positions in Volga-Baku—a 1912 chartered company with Emil Nobel on the board—the Mantashev, Moskva-Kavkaz, and Votyeto companies, Brothers Mirzoiev, Aramazd, Alhan Yurt, Nefterazd, and the Anglo-Russian Maximov Oil Company in London as well as G. M. Lianosov and Sons. Of the three hundred thousand outstanding shares of Lianosov Nobel owned a third in its own name and another twelve thousand indirectly. It also bought outright the transport company Vostotinoye Obschestov Tovarnych Skladov (VOTS) with its network of depots along the Volga and a large fleet of tank trucks. Kama Shipping Company with thirteen tugs and more than fifty barges on the Volga was another purchase.

Nobel's multimillion-ruble investment in other oil producers included the Cheleken-Dagestan Company with fields on the island first staked out by Robert, important holdings in Grozny, the Chimion and Santo firm started by the pioneer Ragosin which included wells and refineries north of Tiflis and a refinery on the upper Volga for lubricating oils, Rapid and Kolschida which operated in Grozny, the I. V. Ragosin Company, and Runo, a landholding group with important properties in Baku—Nobel, Mazut, and RGO each had a third interest. In Emba Kaspiyskoye, organized to develop the new Ural fields, Nobel had a 42.5-percent interest.[9]

Royal Dutch-Shell had beaten Nobel to the Urals by a few months but

by the late summer of 1912 the Swede Wannebo and the Russian Kusne-
zov were organizing extensive drilling and pipe-laying operations, working
the Nobel parcels selected by company geologist Fegraeus. In a climate
and setting even less hospitable to exploitation than Baku the pipelines
were laid, food and supplies brought in from Astrakhan and transported
by camel to Dossor. Huge pits were dug and filled with winter snow to
provide water in the summer. Docks were built at Rakusha on the Caspian. As soon as the oil started flowing refineries were added. The finished
product was shipped by barge directly to Astrakhan. Both Nobel and
Royal Dutch had ambitious plans for future development and by 1916 the
Emba fields were already the third most productive in Russia. Baku led
with 7.5 million tons; Grozny yielded 1.66 million tons, followed by Emba's two hundred forty-five thousand, Holy Island with one hundred
twenty thousand, Cheleken with forty-eight thousand, Maikop and Fergana with thirty-two thousand tons each.

From their large new office building across the Neva on the Admiralty
side of the city, a block from the Nevsky Prospekt and astride the Ekater-
insky Canal, Nobel Brothers by 1916 directed an enterprise which owned,
controlled, or had substantial interest in companies employing fifty thousand workers and producing a third of all Russian crude oil, 40 percent of
all refined, and supplying almost two-thirds of domestic consumption.
Tank farms and oil depots flying the Nobel flag exceeded four hundred in
number, covered more than fifteen hundred acres, and had a storage
capacity of 3.5 million tons. Domination of transportation was equally impressive with the company surpassing its own records, even with the
decrease in overall production. Of all the black gold moving from Baku to
Astrakhan and points north, east, and west in its first thirty years of
business, Nobel shipped more than a third of the kerosene, more than a
fourth of all the fuel oil, and more than half the lubricants.

The expansion and modernization of the Nobel fleet, the investment in
RGO and other Russian companies, the increased expenditures on internal operations, required larger cash outlays than at any previous time in
the company's history. There were regular bank loans and floating of
bonds but much of the financing was done within the company with new
issues of stock in 1911 and 1916 totaling 30 million rubles. There were also
record dividends, especially during the war years. In 1914 the rate was 25
percent; this was raised to 30 percent the following year, and to 40 percent
in 1916. Wartime inflation, of course, influenced these payouts but even

with a 50-percent drop in the value of the ruble the 1916 profit of 75 million rubles on a share capital of 45 million was remarkable.

Uncle Alfred would have applauded the performance, the new issues of stock, the coup against RGO, but none of this was Ludwig's kind of war; it was not his way of winning. He would have found wanting—later in the decade after those great achievements in marine transport—those bold and imaginative innovations which had characterized the company ever since he first had charged onto the scene. Outside stimulation was needed. When Deterding's experts arrived they were surprised at the state of the Russian industry.

Years earlier it had been the Russians who were sent to the Indies to give advice and it had been the Americans, the British, and the Germans who had made the long pilgrimage to see the latest from Nobel Brothers. But the Dutch engineers found a general stagnation, a stultifying self-satisfaction with what they thought to be old-fashioned. No basic research was being done on the use of preheaters or heat exchangers, on regeneration of used sulphuric acid. There was no research under way on the new cracking methods of distillation essential for the refining of gasoline, but instead a reliance on that once far-sighted but now dated Nobel system of continuous distillation. Royal Dutch, with Deterding providing full and enthusiastic support, was installing in Russia the new American Trumble system of distillation, a total departure from the continuous-bench method and especially well-suited to the relatively heavy Russian crude.

Nobel engineers and lab technicians were interested in the process and probably would have experimented with it had the war not disrupted all their plans and programs. There was no time for anything but the war effort. The economy had to be geared to government needs. Emanuel received the orders during the first days of mobilization. Increase production! Send more fuel oil! Ship as much as possible to Petersburg, now cut off from English coal. Do not use the railroads! Ship on the waterways! Hagelin bought and chartered every available barge capable of maneuvering that Marinsky Canal route he had pioneered and ordered the local depot chiefs along the upper Volga to expand their storage facilities. The factories in Petersburg had to be kept running at full capacity.

In Baku Nobel built the country's first factory for production of toluol, that chemical by-product of gasoline distillation essential for manufacture of trinitrotoluene (TNT). The British were obtaining their toluol from coal tar and the Germans from Borneo crude, especially high in heavy hydrocarbons yielding up to 10-percent toluol. Krupp had built Germany's first plant, supplied from Borneo on Shell tankers. Shell later built its own factory in Rotterdam but the Russians continued to import all their needs from Germany—until 1914 when the Nobelites rushed to put up their own plant.

The timing, the rush, the lateness of the hour were typical of the period. Nobel engineers, technicians, and administrators were constantly pushed to the limit to keep up with the demands of their growing company, the demands of an expanding domestic economy, and then of a wartime economy. The problem was a familiar one: shortage of qualified personnel. The chief executives were saddled with greater and heavier loads. In addition to his tasks overseeing Baku and the transportation network Hagelin was assigned management of the export trade and supervision of the several foreign marketing agreements when Ernst Grube, Olsen's handpicked successor, left the company. Deterding hired him away to take charge of the Mazut Corporation; the Dutchman apparently did not care for the Pollack brothers and wanted an experienced outsider to be his man in the company he now controlled. Gustav Eklund, the Baku chief who had to be relieved during the 1905 revolution, was another valuable Nobelite recruited by Royal Dutch–Shell—he was put in charge of their Emba field.

With no one trained to replace Grube, Hagelin took over his desk and became Nobel representative at meetings of EPU, BP, SAIC, as well as delegate to international gatherings to discuss trade and marketing arrangements. It was at one of those meetings in Berlin that Deutsche Bank President von Gwinner called him aside and confided, "We'll soon have war with Russia." When a surprised Hagelin asked the reason the German declared that "our trade agreement is about to expire and we want a better one but Russia won't give us one."[10] That was in the summer of 1914, a few weeks before the guns of August fired their first salvos.

By then Hagelin had some reenforcements in Petersburg. Knut Littorin was brought up from Moscow to take over the sales department and Artur Lessner moved from Baku to sit on the executive board with responsibility for all oil production. Emanuel's youngest brother Gustav replaced him in Baku; he had been working in the various sections of the Baku complex and had spent a few years with Littorin in Moscow learning domestic sales. A graduate of the Cologne Commercial Academy and Emanuel's heir apparent, Gustav was appointed to the company executive council with overlapping responsibilities for financial affairs, production, transportation, domestic sales. His broad background and varied experience on the job, paired with an innate intelligence, promised well for the future of Nobel Brothers.

But from the moment the tsar signed the order for general mobilization in 1914 that future was in doubt. The disastrous military defeats, the steady decay of a regime consumed by the cancers of corruption and incompetence, would lead to paralysis and collapse, to the chaos and terror of revolution and civil war. The future was being determined by forces neither Gustav nor Émanuel nor the thousands of Nobelites could comprehend or vaguely hope to control.

CHAPTER FIFTEEN

End of an Empire

Fifteen days after the mobilization order was issued Russian forces moved against East Prussia. Tragically unready to launch an attack and pitifully ill prepared to endure the sustained strains of modern warfare despite their newfound patriotism and outpouring of support and loyalty to Nicholas, despite the bravery of the front-line contingents, the tsar's troops were swallowed by the forests of East Prussia, decimated by superior artillery, outwitted by superior reconnaissance, outmaneuvered by superior generalship, and ultimately destroyed by the superior self-discipline of the German forces. The fatal weaknesses were apparent before a single shot was fired: a confused self-contradictory command, inefficient supply and transport, ignorance of enemy capability and intentions. The Germans knew that artillery and ammunition were essential to victory but the Russians, forgetting what they had been taught so bitterly by the Japanese, would have to wait until another war and another autocrat's decree that "artillery is the god of war" before making changes. The government had spent eight years working out an arms program. Three weeks before general mobilization they put it into effect. It was scheduled to be completed in seven years.

On the southern front the Russians routed the Austrians, swept through Galicia, conquered the Carpathians. The outcome of the war and the fate of the nation might have been different had they concentrated on this area and remained on the defensive in the north. After the initial brilliant bursts the usual problems and deficiencies crippled the Southern Command and the first sweet victories turned into bitter defeat. By the summer of 1915 the army had lost four million men, most of its artillery, and vast stores of supplies. The enemy occupied Poland, Vilna, Brest-Litovsk, as legions of untrained and unarmed recruits were rushed to the front and sacrificed to advancing German armies.

In Petersburg, as news of the disasters filtered back from the front, the war spirit flagged amidst cries of scandals and increasing shortages. Minister of War Sukhomlinov, the cavalry general who boasted that he had not read a military textbook in thirty-five years, finally was dismissed in June. His incredible incompetence could have been possible only in a government as bankrupt as that of Nicholas II. The critical lack of ammunition, the inability to institute vigorous manufacturing programs to make up the deficits in supply despite the individual proposals of firms like Nobel, were abuses heard by an incredulous Duma which made Sukhomlinov stand trial for his crimes. Their own Committee of War Industries for the Supply of Munitions worked valiantly to ameliorate the situation and the British and French allies shipped in additional supplies. The Entente for a time considered a direct link to their Russian ally across Sweden and the Baltic—at the same time the Germans were trying to persuade the Swedes to join the alliance against Russia—but Churchill's swing to the south at Gallipoli and the Swedish determination to remain neutral blunted both schemes.

The Russian home front managed for a time to meet the demands, to supply their military, and to fuse the spirit of determination with the courage of the fighting men facing the enemy. It was an especially important resolve in Petersburg which produced almost as much ammunition as the rest of the country combined. The Nobel factory responded by working three shifts to increase production of grenades, shell casings, fuses. Additional workers were hired, the clubhouse—with Marta as chief surgeon—was converted to a hundred-fifty-bed hospital to care for the sick and wounded returning from the front in numbers far too great for the government to handle. Emanuel, Hagelin, and other directors funded other hospitals.

But it was too late. The crimes of omission during the twenty years of the reign of a weak-minded tsar ruled by a narrow-minded, iron-willed, and mystical tsarina weighted down the empire with a diseased legacy too great to overcome. The outcome was certain when Nicholas himself took command of the army. With the Tsar of All the Russias at the front Alexandra and Rasputin ruled the nation. The mad mouthings of a malicious monk and the autocratic, blinded visions of an embittered, all-powerful woman molded a tragic and medieval union that could only lead to disaster and total decay of all order, reason, and purpose.

By the first days of March 1917 the rumblings of discontent among the masses of draftees and dissatisfied workmen, reenforced by a hungry citizenry angered by the shortage of food, grew to avalanche proportions as the people took to the streets. Hagelin, moving between the Nobel offices and his home on Liteiny Prospekt, kept a diary of the events.[1]

Thursday, March 8 Streetcars on the Nevsky stopped running. Strikes in several factories. The police clear the streets of demonstrators shouting "Bread!"

Friday, March 9 Only a few streetcars running; others forced to halt. Strikes are spreading. Confrontations between police and the people, mainly on the Nevsky. That street cleared and closed off from the crowds—vehicles and single walkers only allowed to pass.

Saturday, March 10 Streetcar traffic completely stopped. Police, often disguised as soldiers and Cossacks, hold the crowds in check. Dead and wounded lying on the streets.

Sunday, March 11 Quiet in the morning. Again today I could go to the office and work. Stores have boarded up their windows and doors with extra reenforcement. More action in the afternoon. The police were occasionally shooting but the Cossacks still using only whips. A police officer killed by a soldier. In the evening a lot of shooting, mainly near the Nikolai station. Drivers who are carting off the corpses, talk about hundreds. Went to Goukasovs' for dinner but had to make a long detour to get there. His car, bringing his son's wife to the dinner, was stopped several times and had the back window smashed in by an empty bottle. Littorin could not get over the Sampsonievsky Bridge. Picked him up in my car but had to take a long detour. The papers published for the last time.

Monday, March 12 The drivers who took away the dead bodies report that the people have now sworn to come out in the streets armed. . . . soldiers attack then go over to the side of the crowd. By lunchtime nearly all offices and workshops are closed. We let anyone go home who wants to. Lot of shooting on Liteiny. Mrs. Beliamin's nephew was killed by a stray bullet—hit him right in the forehead as he was standing in a window watching the activities on the street. Emanuel, engineer Borggren, who had come on ASEA business, and I couldn't get home. We spent the night with Istomin, the board's secretary, who lives in our office building. Late at night when it calmed down, we went out and walked by the French Theatre. Two young soldiers, hungry and the worse for wear, were sleeping in a nearby courtyard. They had no idea where their company was. My son-in-law could not get over the Liteiny Bridge or even the Troizsky. . . . in the evening we learn by phone that the Duma did not comply with the order to dis-

solve but appointed a temporary committee to try to bring some order out of
the chaos.

Tuesday, March 13 Up early. Telephone call from home reporting all quiet
on Liteiny. Borggren and I went home at half past seven, bathed, ate
breakfast and returned to the office. As we were walking past the south side
of Mars Field, came under fire, shooting at us from both sides. Gruesome!
Borggren threw himself on the ground but I ran on, to get out of the line of
fire as quickly as possible. In the office I found Emanuel, Istomin, my
secretary Werner and Baron Mannerheim, who had just returned from a
visit to Finland and was en route back to the front. He had met Emanuel at
the Hotel Europe where a porter told him to get out as fast as possible
"because they are going around looking for officers." Emanuel took him to
the office and sent a messenger to fetch some clothes for him. He had just
finished changing when we arrived. We all had tea together, then Emanuel
and the General departed. He said "If I can just get over to the Vyborg side,
everything will be all right." Emanuel and Mannerheim were accompanied
by our French representative, Eugène Beaux, just back from his military service. When the three of them were in the middle of Mars Field, a soldier
suddenly came running after them. "There's an officer among you," he
shouted. "Yes, I am an officer," said Beaux and showed the soldier his
French military identification. The soldier saluted and disappeared.

A little while later I left the office and tried to get home but couldn't get
further than the corner of Nevsky. There was so much shooting I couldn't
think of continuing. There's a battle on at the Admiralty: they believe the
police prefect is hiding there. A single issue of a paper came out, "The Social
Democratic Party's Announcement." The Tsar is expected home this evening. Circulars have come out exhorting the people to remain calm, promising them solution of the food problems and a constitutional convention.

There are rumors that three ministers and a former police prefect have been
arrested and are being held in the Duma. In the Duma gather all those who
want to show their support and to the Duma are taken all those they believe
are opposed to it.

In the evening we do not use any rooms facing the street. Sat mainly in the
dining room. Had two searches. They were looking for a machine gunner
who is firing on the street below. They found him on a neighboring roof.
Another machine gunner was found in the tower of the Sergievska Church.

Arrested police officers are driven past us in cars with soldiers armed to the
teeth. A truck comes by with a woman in velvet and furs arm in arm with
happy revolutionaries. My car and driver are loaned to the Red Cross, com-

mandeered by a student who promised it would be used only to transport
bread and the wounded. Between nine and ten at night I opened the door to
my study just slightly, lighting the front window for a second. Machine-gun
fire sprayed the entire facade of the building, firing through the windows.
Because we are on the third floor, only some pictures and chandeliers are
hit. A shocking, frightening experience. In the evening we bedded down on
the floor of the dining room.

Wednesday, March 14 After yesterday's bombardment it was calm until two
o'clock then machine-gun fire again. More house searches, in the attics and
on the rooftops. Prince Gagarin, who lives on the floor below us, is arrested and taken away to the Duma. Rumors about a great army nearing
Petersburg, that the Tsar will abdicate at four o'clock, that the front and
General Ruzsky approve of the revolution, while Minister Frederiks wants
to take Petersburg by force. The Tsarina has asked to meet Rodzyanko, an
ardent opponent but not a socialist. It's reported that the same thing is
happening in Moscow, Tsaritsyn, and Kiev.

Trotsky Senutovich, a member of our board and head of a large cold-storage
plant, went to the Duma to offer his services to organize the food supply
with the aid of others from trade and industry.

They are trying to occupy the Admiralty, the Anitschkov Palace, and the
State Bank. Now there are many more officers among the rebel troops. All
units who march past us are led by officers. Some have marching bands.
The Semenovsky Regiment files past. Everyone and everybody is going to
the Duma—the route goes right past our house. With a general and flags in
front, now with red rosettes. There are some difficulties with the car: some
people want to requisition it but my driver showed them yesterday's cer-
tificate that he was working for the Red Cross and they let him go. Two
more house searches, very polite. Welitchko sent a message that I have to be
in the office tomorrow morning at half past nine in order to organize the
transport program. Also that the worst is probably over.

Thursday, March 15 Four o'clock in the morning and the streets are empty.
At half past seven fewer than usual. It's cold. Twelve degrees. Streets are
cleared of snow and dirt, no soldiers in sight . . . day is quiet. A government
has been formed together with the "soldier and worker council." Temporary
regent will be the Tsar's brother Michael Alexandrovich . . . constitutional
convention no later than three months. The "young" are still not satisfied.
The constitutional convention must *immediately* proclaim the democratic
republic. They hold meetings, arrange processions, and work with all their
might. Several government bureaus are working normally. The price of
butter and eggs is dropping. . . . the soldiers are more disciplined, but the
workers' organizations are demanding control of the military. They declare

themselves willing to obey if the orders are not in conflict with the "soldier and worker councils."

Rumors of the arrival from the front of Nikolai Nikolaevich. Prince Tschaikovsky is arrested . . . many former ministers as well. Protopopov turns himself in. Sukhomlinov is in danger of being murdered—lynched. Emanuel arrived at seven o'clock, had walked home, was exhausted. We managed to get him to bed at nine o'clock.

Friday, March 16 Sunny but cold. Going to send a telegram "All right" home to Sweden. No information from Baku except a business telegram from the 13th. Engineer Strichov arrived from Grozny—the trip was as difficult as it has been all during the war, but especially rough in Moscow. . . . The day is calm, only parades and meetings with demands for a democratic republic. The Tsar has abdicated in favor of his brother who is refusing to accept it but at the same time offering to serve the people. Organization is better in food distribution and the government council is functioning.

Arrests are continuing, among them Count Kokovsev and General Rennenkampf. . . . Rumors that Kaiser Wilhelm has been murdered, that the garrison in Vyborg is marching on Petersburg, that General Ivanov is approaching from the south. But so far no one has arrived and in Ivanov's army four regiments have gone over to the "people." There's at least one accurate report among all the rumors: Finland's Governor General Zein has been arrested and Finland has declared its independence.

In Kronstadt some hideous happenings. Admiral Viren murdered, masses of naval officers tortured, thrown into furnaces, drowned, or otherwise slaughtered.

After the Tsar's abdication the local authorities, who have thus far been opposed, go over to the new powers—very quietly and peacefully. Can now resume the struggle to end the conflict between "workers and capitalists." I wonder if it will go just as calmly. Emanuel did not come over this evening—he's probably had enough of being forced to go to bed as early as nine.

Saturday, March 17 Ten degrees and snow. Today we the oil industrialists are supposed to coordinate our interests and organize the transport of our products over the entire country. . . . during the war the state was in charge. I was personally granted permission to use my own car. . . . Snowstorm.

One month later, compliments of the German General Staff, Lenin arrived at the Finland Station. For the next months his Bolsheviks outshouted, outorganized, and outmaneuvered their divided opponents, rallying the people with cries of "Peace! Land! Bread!"—forcing the center of political gravity far to the left and eventually seizing all initiative from the Provisional Government. Lenin repeatedly demonstrated that he was the most adaptable, decisive, imaginative, and single-minded leader on the scene but the ruthlessness that would follow absolute control was not yet apparent, least of all to the many foreign observers and visiting socialists like Hjalmar Branting, Sweden's first Social Democratic member of parliament and later the organizer of its first Social Democratic government. Along with numerous other European liberals he enthusiastically applauded the downfall of the Russian government and the rise of Lenin.[2]

The anarchy in Petersburg soon spread to other cities. The collapse of the imperial government and the abdication of the tsar, the only unifying forces binding together a vast and diverse land, left disorder and confusion in its wake. Estates were pillaged, peasants seized the land, troops deserted by the thousands in order to return to their villages and share in the spoils. As the workers' demands became more radical, the shortages of supplies and raw materials critical, and the continuing pressures of the war overwhelming, production in the factories drastically declined. The transport system, crushed by the thousands trying to flee or to join the chaos, approached total breakdown as engines and cars were neither repaired nor replaced. To prevent mass migration to other lands the government decreed that no more than 500 rubles a month could be taken or sent out of the country, but at this stage most of those fleeing were non-Russians.

At the beginning of the war there had been more than two million Europeans living in Russia—a quarter of a million Swedes and six times as many Germans, the two largest groups. As their stream of escape turned into a flood the many Finns in the empire returned to their homeland, now struggling for its own independence. The wives and children of Nobel's Swedish employees moved back to the safe shores of their neutral country but almost all the Swedish salaried and wage-earning personnel stayed on. Their future well-being depended on the future stability of the country, on the survival of the factory, the recovery of the oil industry. Not until their lives were threatened did they leave.

The refugees in Sweden could read their countryman's new book *How to Do Business with Russia*. Written just prior to the revolution this practical manual by a Swedish veteran of twenty-seven-years' experience selling machinery all over the empire confidently predicted that "whatever changes do take place, they will be mostly for the best." Russia, the manual declared, "after many centuries of oppression and repression, will be opened out to the capital, enterprise and energy of the nations of the West," and "the numbing and deadening power of the 'Tchenovniks' [bureaucrats] and of the old corrupt police, who levied blackmail on the merchants and manufacturers, will now be curtailed."[3]

Any residue of such misplaced optimism was quickly dispelled by the events of November. The guns of the Peter and Paul fortress fired on the Winter Palace in concert with those of the cruiser *Aurora* and the disorganized, ineffectual Provisional Government, sterilized by its inability to bring the peace so desperately needed, collapsed. Lenin and his Bolsheviks seized power.

They wasted no time showing the nation and the world the meaning of totalitarian revolution, relying on old regime methods to stamp out all opposition and enforce their own iron rule. All newspapers but their own were suppressed, editors imprisoned, and the only printed or spoken word permitted was that of the Bolsheviks. Their avalanche of decrees covered every phase of economic, political, spiritual, and intellectual life. Just as reformed poachers make the best gamekeepers, so did the Soviets know best how to control the dissidents and crush all other revolutionaries and other rumblings of opposition.

The workers' and soldiers' committees, the soviets, were soon demanding great sums of money from industrialists and putting their representatives on managerial and administrative staffs. Short of fuel, raw materials, and supplies of all kinds, Nobel, Lessner, Renault, and most of the other factories succumbed to the chaos and closed down. All Volga transport was nationalized but that only worsened the situation. Then the factories were nationalized and the owners and managers dismissed. Salaries and pensions were terminated, bank deposits seized, and the ordinary bourgeois permitted to draw only 150 rubles a week. Prices skyrocketed: a sack of flour cost 800 rubles, a pood of sugar 1,000, gasoline 50 rubles a pood—before the war it had been .2 ruble—but only the authorities had vehicles anyway. A series of shops were opened where jewelry and other valuables of the dispossessed could be sold and the bourgeoisie took to the streets to perform any kind of labor to keep from starving.

Bread, when available, was rationed at forty-five grams a day, and as famine threatened in the cities there was an increase of violence with widespread pillaging of homes and buildings suspected of containing hoards of food and spirits.[4] Street violence soon turned to official terrorism as tsar and family were murdered; the Alexander Nevsky monastery was raided and occupied; there were mass slayings of officers and executions of bourgeoisie, of Socialist Revolutionaries, of Cadets. The revolutionary turmoil and random instances of cruelty in the fervor of the struggle were replaced by an official policy of terror which was soon institutionalized as an integral part of the Soviet system of government. Lenin advised his followers not to whine over the useless vermin blocking their path.

In Baku the vermin blocked the way for some time after Lenin and his Cheka—his secret police—had cleared a path in Petersburg and his new capital of Moscow. A Bolshevik stronghold in the heart of Menshevik-dominated Transcaucasia, Baku was the pearl of southern Russia prized by the crumbling empires of the Ottoman Turks, the tsars, and the kaiser, and needed desperately by the new empire builders the Communists—but also coveted by the British and their allies. Hagelin went to the area as soon after the March revolution as conditions permitted and found the city far less chaotic than the north. With representatives of the other companies he met with leaders of the workers' organizations and negotiated the necessary compromises, settling outstanding differences between the two camps. The fact that he did not have to deal with a soviet, with one of the worker-soldier councils, was much to his liking and advantage for when he went on to Astrakhan he learned that the ground rules included confrontation with just such a council.

The trip to Baku was Hagelin's last. Fittingly he made his final crossing of the Caspian on board the tanker *K. W. Hagelin*. It was a restful introduction to the events in Astrakhan where the council presented him with a long list of demands: higher wages, shorter hours, direct participation in company affairs, voting rights in management of workshops and on board ships, and distribution of profits. Discussion between Hagelin and the workers' leaders was long and occasionally heated but no compromise was forthcoming. The machine shop was shut down and Hagelin was summoned to an open meeting of several hundred workers.

It was in fact a trial with the Nobel director accused of being a capitalist incapable of comprehending the position of the workers or recognizing their interests. The main orator, the prosecutor, was a recently hired employee whom Hagelin did not know but suspected of being a professional agitator skilled in rousing a crowd and in suppressing any who stood in opposition. At one point a group of the older workers gathered around their Karl Vasilievich to protect him from possible seizure by those whipped into an angry frenzy by the shoutings of the prosecutor. Compromise, however, eventually was reached as calmer heads prevailed. An eight-man committee was formed to arbitrate the issues. Representing the Nobel Company were Hagelin, his Astrakhan manager, his old friend Alexander Merkulyev, and the head of the Caucasus-Mercury line. After four days of meetings the final report was made and accepted by the workers. But not all of them were satisfied; those who had protected him at the meeting came to his room to warn of an attempt on his life. Hagelin sailed for Saratov the next day. He was not a moment too soon. The news of Nobel Brothers' record 40-percent dividend was just being announced in Petersburg.

It was the last good news the company had to report. Hagelin's experience in Astrakhan was repeated over and over again as the revolutionary councils braced their once-powerful employers with demands. Orders flowed from Moscow in a staccato procession of proclamations, seizing individual Nobel installations. Gustav, brought up from Baku to take charge after the March revolution, was called to endless meetings and did what little he could to forestall the inevitable. Emanuel's effectiveness as diplomat, financier, and director was abruptly terminated when the old order collapsed, when the government offices and banking houses and business firms were seized by a new breed of Russian. He turned negotiations and direction over to his younger brother. On the first day of August 1918 Gustav and the other oil industrialists were summoned to a Moscow conference. As one of the participants put it, "They're asking us to arrange our own fourth-class funeral, one in which the corpse himself drives the hearse."[5] The petroleum industry was nationalized.

It was War Communism, total centralization and control of the economic life of a country now declared to be a military camp. An overture was made to Hagelin to take over technical leadership of the industry—he had impeccable credentials: proletariat-peasant background, a worker who had risen through the ranks, understood and respected by the workers. But Hagelin declined. He had absolutely no political sympathies with Russia's new masters and he had no desire to be subject to the whims, inexperience, and stupidities of committees of workers who would watch his every move.[6] The offer was an interesting indication of Bolshevik regard and fifteen years later the viewpoint was confirmed by a favorable portrayal of Karl Vasilievich in Soviet writer Alexis Tolstoy's novel *Nafta*, a pedestrian propaganda piece on Baku, the industry, and the local revolutionaries.

At the time of Hagelin's visit to Baku after the March revolution a

Moslem government was in power. It was just getting organized when Russian army units seized the capital and installed the Bolsheviks, sending the Moslems packing to the ancient Azerbaidzhan capital of Elizavetpol. Thousands of the remaining Tatars—eighteen thousand according to one Iranian reporter—were massacred by the Armenians who took a belated bloody revenge for the slayings of 1905.[7] As blood again ran through the streets and winding alleys of the city the Council of Peoples' Commissars nationalized the oil properties and the town was soon in chaos, encircled and blockaded by rebel tribes who prevented food and supplies from entering.

To the hunger, the endless parades and demonstrations of workers and soldiers, were added the absurdities that so often characterize the early stages of revolt and establishment of new and inexperienced authority. A former insane-asylum inmate was named Minister of War and promptly proposed the election of a donkey to the council to represent oppressed animals; an illiterate sailor was placed in charge of the schools, and a well-known pimp was given control of public welfare.[8] The car, home, and office of the Nobel chief—Lessner went back to Baku to replace Gustav—were requisitioned and along with most of the senior personnel Lessner was imprisoned when he refused to make ransom payments to the council. In the refineries and oil fields Nobel's section chiefs were put under the command of council-appointed watchdogs as normal work schedules proved impossible to maintain; discipline and routine were replaced by an endless series of meetings, decrees, and denunciations.

The more Lessner resisted the demands of the council the greater his difficulty. He was denounced finally as a counterrevolutionary and sentenced to death. The Bolsheviks commuted this to banishment, but no sooner had he reached Petersburg than the Baku government was overthrown by a coalition of Socialist Revolutionaries and Armenian nationalists propped up by a small British force. The British had been sent there to save the oil for the Allies and to prevent the Germans—already in command of the Ukraine and the Black Sea—or the Turks—who had regained Batum—from occupying the Apscheron peninsula. But the small British-Russian-Armenian force was no match for the Turks and when they ordered an artillery bombardment of the town the British hastily withdrew to Persia, leaving Baku's Armenians to the mercy of Azerbaidzhani troops eager to revenge the March bloodletting. The Turkish force remained outside the city, honoring that ancient Oriental custom which declared "the gold, the lives and the women of the enemy" to be the property of the victors—the Tatar troops—for a period of three days. At the end of that traditional time a decree was issued forbidding further pillage and slaughter. Turkish troops erected gallows on the streetcorners, hanging fifty who violated the ban the first day, twenty the next. The town soon returned to normal.[9]

As in 1905 Nobel installations escaped serious damage. With few Armenian employees and a good reputation among the Tatars there was no wholesale invasion or vandalism. But the rampaging hordes were not about to let Armenians escape merely because they were seeking shelter with Nobel—three were seized inside Villa Petrolea.

For a few months there was calm, a return to nearly normal conditions although Baku was never completely free of violence and bloodshed. A performance of a well-known Azerbaidzhani operetta for the former Persian ambassador to Petersburg was interrupted by an argument in the audience. Shooting broke out—most of the audience carried weapons—and a dozen or so were killed or wounded before order was restored and the management, with due apologies for the interruption, continued the show.[10]

The calm in Baku stimulated the hope in Petersburg that reconstruction and revitalization was possible. The industry had been nationalized by Moscow decree, but the Bolsheviks were not in control of Baku or Batum. Lessner was back at his post and Hagelin decided to see for himself what was happening. With the Caspian closed to all traffic, the rail lines disrupted, the cities sealed off by civil war or Bolshevik confiscations, he had to take the long way around through Poland, the Ukraine, and the Caucasus—German-occupied territory. His trip started in Berlin in the offices of Deutsche Bank President von Gwinner who arranged introductions for his old friend and business colleague to the Department of War and the Ministry of Foreign Affairs. But Hagelin got no further than Sevastopol when he learned that the Turkish troops, their government and empire in total collapse, had to withdraw from Baku and the Caucasus.

Frustrated in his plans he had to return to Berlin, content with his inspections of Nobel installations in Warsaw—they had a deficit with Petersburg of 9 million rubles—and in Kiev where the chiefs from Kharkov, Rostov, and Odessa gathered to report on developments. Hagelin instructed them to audit their books, make the yearly accounting, and put any profits or excess cash into purchase of houses near their cities. The order was never carried out. Three weeks later the armistice was signed and the Germans withdrew from the eastern front. The Turks and Germans departed; the British rushed in to fill the vacuum hoping to save Baku and Transcaucasia from the Bolsheviks.

Working with the government of the Azerbaidzhan Republic the British representatives were judged by the Nobelites to be less capable than their Turkish predecessors and, much to Lessner's annoyance, consistently sided with the government against the owners and industrialists. Was it to gain the loyalty of the workers, he wondered, or perhaps to force a wedge that could later be used to strengthen the British presence in the area, perhaps to replace Nobel?

Work in the fields slowed to a lazy and largely unproductive pace as material shortages were added to the other problems. Storage tanks and reservoirs were filled to overflowing. The Caspian had been closed ever since Lessner had returned on the last ship to Baku in August 1918; the flow to Batum by rail and pipeline was reduced to a trickle. He started building additional storage facilities at Bejouk-Schor but before they could be completed a Soviet armored train led a surprise attack on its sister republic, forcing the collapse of the Azerbaidzhan government. The British, who had retreated to Batum, prepared to evacuate the entire area. With Baku in Bolshevik hands no oil would be coming to Batum; but more important, the British prime minister had just signed a trade agreement with the Soviets pledging to refrain from all anti-Soviet activities in any former tsarist territory.

It was the former Baku businessman Leonid Krassin who negotiated the agreement with Lloyd George and another veteran of the land of blood and oil who had sealed the fate of the world's first Moslem republic and then crushed the sister socialist republic in his native Georgia. But who would know better than the Commissar of Nationalities, Lenin's chief of staff for Caucasian policy, how to convert neighboring countries to Communism by the force and threat of armed might? And who could better appreciate Stalin's use of bullets and bayonets than a twenty-one-year-old Bolshevik, a poor son of Black Sea peasants? Lavrenti Beria, student at the Baku Technical College, watched it all. By the time Stalin completed the conquest of Georgia Beria was a member of the Caucasian Cheka, starting along that gruesome road which would lead to his total control of the Caucasus through the 1930s and even greater power in the 1940s.

For Stalin and finally Lenin, who was slow to agree, there was no other choice. With Finland and Poland already torn from the Russian corpse the Soviets could not afford to lose their southern flank. The Caucasus was too important a source of raw materials, of oil, of food for the new Soviet Republic to let it slip from their grasp. And Stalin recognized the strategic importance of this bridge between Russia and Turkey, this key to Asia. There was at least one Bolshevik leader who had not forgotten the dream of the tsars, the yearning of Catherine the Great for a southern port.

The end for Baku was surprisingly swift. The Soviet Eleventh Army under the command of Tukhachevsky with Sergei Kirov as his commissar trapped twenty-two thousand of Denikin's troops at Novorossiisk as they were attempting to board British and French ships. A few weeks later, on 28 April 1920 after the usual protestations of eternal friendship and denials of any aggressive intention, the Soviet armored train roared into White Town. Kirov and a young Armenian, Anastas Mikoyan, were on board. By the end of the year Mikoyan's native province was also incorporated into the Soviet state.

There was no resistance. The Armenians and Azerbaidzhanis proved more docile than the Georgians. When the Red Army invaded that republic in February 1921 the ragtail and ill-equipped Georgians resisted. For more than a month as the conquering Bolsheviks let loose their forces on a rampage of murder and mayhem the Georgians and their Menshevik Social Democratic government stubbornly refused to be beaten. But Lenin and Stalin continued to talk about popular uprisings, the right of self-determination, the liberation of the Georgian workers.

With the Bolsheviks back in Baku and no Turkish, German, or British threat to their presence, there was no effective opposition to their plans of reorganizing society and taking over the oil industry. Those who did oppose their decrees—enemies of the new order like the director of the local Asov Bank—were taken out and shot. Ministers of the Azerbaidzhan government, despite Soviet guarantees of safety spelled out in the documents of surrender, were put to the sword. Thousands more were arrested and taken to the Cheka prison on the nearby island of Nargen, there to disappear without trace. There was no security for the bourgeoisie, no missed opportunity to make existence in home or office precarious and perilous. The Cheka introduced a form of slaughter new to the area: systematic, routine extermination of masses of people. This was not the wild, bloody rampage of the Armenian-Tatar massacres but orderly, planned executions of any defender of the old order. It was organized, methodical terror.

A "week of plundering" was decreed: the revolutionary proletariat "oppressed and deprived by brutal capitalists of the very necessities of life" were allowed one week to search the homes of the "capitalist bloodsuckers and their parasites" for clothes, money, furniture, utensils. Opposition to the decree, to the search, constituted a revolt against the state and was punishable by death. Homes and offices of executives, engineers, technicians were requisitioned and only those individuals judged essential to the industry were given the option of remaining on the job.[11]

Lessner, with Malm and Wannebo working with him, stayed on another four months. As Malm put it, "there was always the danger of being stood up against the wall and shot," but none of the Nobel chiefs really believed that the danger was more than a passing phenomenon, that the Bolsheviks could continue very long in power. In September 1920 Wannebo got out to Sweden, Malm and Lessner to Europe. Nobel's last chief in Baku, traveling via Tiflis, Batum, and Constantinople, was met at his arrival in Germany by Emanuel and Hagelin who had been out for two years. Karl Vasilievich had made his last trip to Petersburg in the summer of 1918 when he was accredited as a diplomatic courier. As a former consul general he had little trouble arranging official orders at the Swedish Foreign Office.

Emanuel came out about the same time as Hagelin, traveling overland from Yessentuki in the south of Russia to Sassnitz in the north of Germany. Helpless in the face of the Soviet decrees, his factory, corporations, and home taken from him, his oil empire in a shambles, he moved in the spring of 1918 to the little town of Mineralnye Vody in the safety of the northern slope of the Caucasus. This spa on the River Kum, starting point for Solzhenitsyn's novel *August 1914*, with the neighboring towns of Kislovodsk and Yessentuki was the rallying point for many members of the aristocracy and a flock of industrialists. Grand Duchess Vladimir and her sons were in residence; Mantashev was firmly planted in the hills, still giving wild parties but also subsidizing a private guard. Emanuel was there with Gustav's wife and three children. For a time the area was safe; there was a White Army in Stavropol screening the north and a newly established German-supported Georgian Republic to the south. The Red Army was busy elsewhere and there were no serious shortages of food. The only shortage was money. The exile colony was cut off from the rest of Russia and had nothing but what its many residents managed to carry with them—jewels, stocks, bonds, credit notes, silver, valuables of great variety, but very little ready cash. Emanuel proposed and then implemented the solution: the printing of a special currency to be purchased with these possessions. The banks of Kislovodsk readily agreed. The new Nobel notes were printed and the exiles were able to make the conversion.

As summer came and went new exiles arrived; others just as suddenly departed, finding one or another escape route to the west. But the Soviets were solidifying their power, defeating the disorganized White armies. In the south the bigoted, reactionary General Denikin proved to be one of the best allies the Bolsheviks had; he used his armies to blockade both Georgia and Azerbaidzhan as he opposed any effort to coordinate with other forces facing the Soviets in their sweep toward the Caucasus.

Gustav and Emil in Petersburg feared for Emanuel's safety and they dispatched a young officer, son of one of their employees, to help plan an escape. Travel permission never would have been granted a Nobel so forged passports, exit permits, and other documentation had to be procured. Cooperation of a former Baku deputy prefect of police was secured, the exit permits and other papers in false names properly stamped and taken by company courier to Yessentuki. There the documents were delivered to Emanuel, Gustav's wife and her children. Dressed in peasant clothes and avoiding the town of Kislovodsk where Emanuel was too well known the party drove over the steppes directly to Stavropol, still under the control of the White Army—Emanuel knew the old general in command. They were stopped just once by a Red Army patrol. A few hours after they left by train for Kiev the Red Army entered Stavropol. In Kiev the group was taken in tow by the local Nobel representative and put on the train to Warsaw where Emanuel by chance met the son of an old business friend in Berlin; it was, in fact, his godson who was then an officer in the German army. He quickly made arrangements for their journey to Berlin and from there Emanuel had no trouble getting to Stockholm.

Gustav and Emil took a more direct route but their departure was delayed by a stay in a Cheka prison. On 30 November they were arrested and taken to Gorochovaya 2 where the local chief, an attractive Jewess reputedly as fond of fancy jewelry as she was of torture, personally interrogated them. With a dozen young men of the Cheka standing in a ring behind her she charged the two Swedes with their crime. The British in Baku had arrested a local Bolshevik in the employ of Nobel Brothers as secretary for engineer Felix Hedman. If the British would release their prisoner the Cheka would release the Nobels. If the British decided to execute their prisoner the Cheka would execute the Nobels by shooting, hanging, or however the British did it—a final bit of insanity climaxing a year of frustration, harassment, confiscation, terror for the Nobels.

The Swedish Legation aggressively intervened in the arrest of two of their citizens and by a judicious combination of flattery, threat, and pleading secured their release on the promise that they would report daily to Cheka headquarters. But in a few days the legation closed down and its members returned home. Gustav and Emil were quick to follow and by train, sled, and foot crossed the border into Finland; from Vyborg they managed the journey without incident to Stockholm, arriving three days before Christmas. The last Nobels had departed. After three-quarters of a century the Swedish saga of a family in Russia came to a bitter end.

CHAPTER SIXTEEN

Krassin and the Communists

But was it really the end? Could the Bolsheviks continue in power? Neither the Russian nor international oil industrialists believed it possible Certainly the Nobels did not imagine they had seen the last of Russia. Somehow and sometime soon they would regain control of what had been taken from them. The total and irrevocable loss of all their property was inconceivable although they knew that certain possessions would never again be seen. When Marta fled to Finland she deposited her jewelry and other valuables in the Volga Bank, believing they would be safer in those vaults than in the Nobel summer home. The Volga was one of the first banks raided by the Bolsheviks. An insignificant loss when measured against the Nobel empire but in retrospect a considerable and needless sacrifice to the Communist treasury.

If the family resigned itself to the loss of personal possessions it was certainly not ready to surrender forever its factory or its command of Nobel Brothers. Its members made preparations for their return, not from Stockholm but from Paris where the émigrés gathered in droves, each more certain than the other that it was only a matter of time before they would be back in Mother Russia. From a small second-floor office, humble contrast to the magnificence of the Petersburg complex, the Nobels made ready.

Emanuel remained the titular head of the firm but, like Rockefeller before him, retired from the daily struggles of the chief executive's chair. He celebrated the occasion and his sixtieth birthday by taking his nephews and the two sons of Hagelin—Emanuel was like a father to them all—on a cruise to the North Cape. On a hill overlooking the world's northernmost city they drank champagne toasts to the health of Uncle Emanuel as the midnight sun blazed over Hammerfest—a typical grand gesture by the bachelor patriarch who loved parties and people.

Returning to Paris he was grateful to hand over the routine and frustration to Gustav and his small staff: brother Emil, Karl Vasilievich, and Ragnar Werner—Gustav's schoolmate and Hagelin's secretary, a capable addition to the group with his fluency in English, German, and French in addition to Swedish and Russian. The Nobels moved into the Hotel Meurice and Hagelin took a little apartment on Avenue Henri Martin. He had first gone back to Sweden but after a battle with the tax authorities—his chief antagonist was later jailed for fraud—he moved to the French capital.

For all the White Russians and deposed oil barons in the years immediately after the revolution, Paris was where the action was. It was certainly not in Stockholm even though that city served as the corporate headquarters of Nobel Brothers—Emanuel and Gustav had managed to transfer titles and papers from Petersburg before the Bolsheviks could lay claim to the international assets of the company. Nobel thus managed to retain its Société Franco-Egyptienne with holdings in France, Egypt, and Syria, and they still had their SAIC marketing outlet. The skeletal remains of EPU was of little use to them but they did make every effort to recapture the two markets that were once Nobel provinces: the newly independent nations of Finland and Poland. Gustav negotiated with the Warsaw Posener Bank which put up half of the 100-million-Polish-marks capitalization to establish an integrated company with drilling leases, refinery and sales outlets. But the relatively small Polish market was not very profitable. What was previously a tiny appendage on the giant Nobel body, one which could be supported without strain, was now an independent and costly unit that the Nobels in exile could ill afford. They let it be known that partners would be welcome.

It didn't take long for the news to reach 26 Broadway. Standard Oil of New Jersey, largest of the Standard offshoots from the 1911 breakup of the company and inheritor of European interests, was the logical candidate for cooperation with Nobel. After a series of meetings in London with Emanuel, Gustav, and Emil shuttling back and forth across the channel an agreement was reached in November 1919. Standard purchased half the Nobel interests establishing POLNOBEL—the Standard-Nobel Company in Poland.[1]

With this cooperative venture successfully launched it was relatively easy to negotiate a similar joint operation in Finland. The primary goal was to secure a steady source of oil and protect a market they had previously dominated, now threatened by Standard affiliates in other

Scandinavian countries. In April 1920 agreement was reached and Nobel-Standard appeared in Finland. Nobel and a group of local businessmen each put up 30 percent of the 30 million Finnish marks, Standard the remaining 40 percent. Hagelin, who had initially developed the Finnish area and in 1911 had chartered a separate Nobel company, was appointed to the board of directors. Price agreements with Shell, the main competitive threat, were arranged. The new company returned a welcome profit to the Nobels for many years.[2]

These agreements in Poland and Finland involved only the economic periphery of the basic mutual interests of both Nobel and Standard: alliance in Russia—echoes of the 1890s and the abortive attempt to divide the world, revival of that dream of Olsen to father a formal merger of the two giants, and confirmation of the continued belief that the Bolsheviks could not retain power, that the oil properties would be returned to their rightful owners. With the rising competition from one of those owners, Henri Deterding, the board of Standard was compelled to make every effort to enter the Russian market, to secure its own footing in the oil-rich lands of the south. Its domination of the American market had been shattered by the 1911 Supreme Court decision and the rise of the American independents—Gulf, Sun, Union, Pure Oil, Texas—and it was increasingly concerned with European developments and the expanding European market.

In January 1919 Standard signed a contract worth a third of a million dollars with the independent government of Azerbaidzhan for the purchase of eleven plots in Baku. When the news came a few months later that the Anglo-Persian Oil Company was considering purchase of Nobel shares the board moved swiftly to make contact. Heinrich von Riedemann, former Standard chief in Hamburg, was the channel selected and he promptly arranged for Emanuel and Hans Olsen, Gustav and Hagelin, to meet with James Moffett, Everit Sadler, director and overseas expert, and Olsen's old friend and confidant Powell from Anglo-American.[3]

On the suggestion from the Swedish side of the table of a possible sale of up to 50 percent of Nobel investments in Russia it was Powell who forwarded the strongest recommendation to New York. "Cooperation at any cost," was his plea; he feared some other company would rush into the breach and close a deal first. Standard wasn't aware of the fact, but this was no danger. Nobel did not consider other offers once serious discussions commenced with Standard. The family preferred to work with the Americans, to have the power of Standard and the backing of its coun-

try with the businesslike approach of its cooperative consular corps helping to hold the Nobel banner. How else could they even hope to withstand
the power of the Bolsheviks who were then sweeping down from the north
to threaten the Moslem Republic of Azerbaidzhan? Association with
Deterding or the British firms would not be nearly as effective; they were
already in the area and deeply involved in their own struggles to regain
their properties. Only Standard could provide fresh reserves of strength for
future battles. Nobel's preference for alliance with Standard was as logical
as it was wise; circumstances demanded it just as historical precedents
permitted it.

Standard's board of directors put their tentative stamp of approval on
the proposal, predicating final assent on the results of an on-scene survey.
Sadler was sent to Baku, traveling by U. S. warship to Batum and the
remainder of the way by rail. His report was as thorough as it was
enthusiastic and his strong recommendations led to the March 1920 opening of serious negotiations with Gustav in Paris. A preliminary agreement
was signed on 12 April. Two weeks later the Red Army captured Baku,
but the discussions continued. Standard showed its faith in the future
of freedom in Russia by advancing a half-million dollars to Nobel to
purchase additional oil-producing lands in Baku for the joint account of
the two companies. Twenty-six Broadway was obviously convinced that
the Bolshevik regime would not last, that one day Standard's engineers
would be working in the area described by Sadler as "a production
engineer's dream" and "a wildcatter's paradise."

In May Gustav and Ragnar Werner sailed to the United States for final
negotiations. For eight weeks they bartered and bargained; discussion occasionally took a dramatic turn and there were some critical moments but
the outcome was never in doubt. Both sides wanted the agreement. It was
only a question of price and procedure. Chairman Teagle and his board
were quite willing to gamble on losing their entire investment against the
unique opportunity of gaining wide and easy access to half the Russian industry by buying half the Nobel family's interests in their company. It was
easily worth the relatively few millions the Nobels were asking.

The problem was physical possession of the paper. Of the hundred forty
thousand shares of common the family and closely affiliated individuals
like Hagelin owned thirty-six thousand shares; but twenty-six thousand of
those were in Russia, in Communist hands. Also, the thirty-six thousand
did not represent majority control; it was therefore decided to spend $5
million to acquire that control by buying additional shares, the costs to be

split by the two parties. Finally certification of ownership was accepted in lieu of actual presentation of shares still in Russia.

The price? Eleven and a half million dollars to be paid through Standard's Swiss holding company, the Schweizerische Handels-und Beteiligungs-Aktiengesellschaft. Organized in 1911 with 1.5 million Swiss francs for the purpose of purchasing Pure Oil properties in Germany, this wholly-owned subsidiary was to pay Nobel over a period of two years the sum of $6,568,000 upon receipt of thirteen thousand shares or certificates. The balance was to be paid upon delivery of an additional five thousand shares and the restoration to Nobel of their Russian properties.

On 30 July the final papers were signed. A month later while Teagle was briefing the secretary of state—who expressed his unofficial satisfaction with the investment that would later represent a tenth of the American claim against the Soviets—a triumphant Gustav returned to Stockholm, his pockets bulging with the guarantee of financial security for the Nobel clan. That was surely one of his goals from the moment he and Emanuel, Olsen and Hagelin first made contact with Standard. Looking back on the performance from the comfortable perspective of time Gustav's achievement was brilliant, a remarkable and profitable masterstroke.

Standard was not alone in purchasing defunct Russian companies, in dealing with destitute and sometimes desperate entrepreneurs. The cafés of Paris, rife with rumors from the homeland, were like brokers' branch offices with securities traded on a curb market and icons, paintings, jewelry, and other treasures changing hands like a Baku bazaar. It was there that Leon Mantashev sold shares in his companies; then his paintings one by one until only "September Morn" remained. But even that had to go and the purchaser was another Armenian—his father's onetime flunky Calouste Gulbenkian. Enriched by his safer investments in the Mideast, "Mr. Five Per Cent" again demonstrated his shrewdness by purchasing from the émigrés their art and not their worthless shares.

Deterding was the master at that. He bought large blocks of stock from Mantashev and many others including Schibaieff, the Atzatourov Company, Baku Russian, and Nikopol-Mariupol, the country's largest pipe manufacturer. The Hollander was planning to build a pipeline all the way from Grozny to the Black Sea port of Novorossiisk.[4] He didn't want art. He wanted oil and he wanted it in massive quantities in order to drive Standard and all other competitors to the wall. The unity and concerted cooperation of the war years when, in Lord Curzon's words, "the Allies floated to victory on a wave of oil" were replaced by a renewal of hostilities in the private battles of the oil companies. Deterding knew that he was taking great risks in purchasing Russian shares, but he also knew he might never again have such a golden buyer's market. He could not ignore the opportunity.

He was also taking vigorous steps to reclaim what had been seized by the Bolsheviks. The Shell representative in London was negotiating with the head of a newly arrived Soviet trade delegation, a prewar millionaire and managing director of the Petersburg branch of the Siemens-Schuckert Company. This was Leonid Krassin, the Count Witte of the Communists, chief Soviet diplomat for negotiations with the west and Commissar of Foreign Trade. He had already passed with honor and glory through the posts of Red Army Minister of Supply, Commissar of Trade and Industry, Commissar of the Petroleum Industry, Commissar for Transport, and had organized a commission to catalogue and save the art treasures of the country and established a Society for the Study of Exact Sciences, a powerful magnet which attracted academicians and intellectuals to the cause. An incredible overachiever, "one of those who felt deeply the poetry of work," one to whom "work is sheer joy and pleasure," as Gorky described him, but also one whose stamina and health were bound to be broken by the strain.[5] The collapse did not come until 1925 and by then he had brilliantly outmaneuvered the oil companies, masterfully playing one group against another, wielding the weapons of trade to shatter western boycotts while stimulating the reconstruction of an oil industry so vital to the government as a source of foreign revenue and as the primary fuel for economic recovery.

Krassin was as flexible in his philosophy as he was in his operational tactics, an outstanding proponent of Lenin's basic diplomatic strategy utilizing Comintern and Foreign Commissariat to exacerbate discord in the enemy camp and relieve the pressures on the Soviet state. He was also an enthusiastic supporter of Lenin's New Economic Policy (NEP), that final desperate reversal of principle and the first of many Soviet posturings to the capitalist west for money, tools, and techniques to save its own regime, to retain the incumbents in power. NEP was formulated at the Tenth Party Congress while the peasants were in revolt against forced requisitions and the Red Army was crushing the armed uprising of sailors and workers at Kronstadt. At the time the Soviets were negotiating a commercial treaty with England, arranging for the exchange of trade missions; that lead was followed in the next three years by most of the European countries which began to extend diplomatic recognition as well. Sweden negotiated its trade agreement and granted recognition in 1924.

It was Krassin who made the first breakthroughs with the Swedes, Krassin who was responsible for forcing the first cracks in the western blockade of Bolshevik Russia. As early as 1920 while en route to his May meetings with Lloyd George and the British cabinet Krassin stopped in Sweden where he signed contracts with more than a dozen firms for the purchase of farm machinery and a supply of telephone and telegraph equipment. For the depression-worried Swedish government and the interested suppliers the contracts were a great boon; for Krassin and the Soviets the success in Sweden was an important precedent to take into talks with the British. The veteran revolutionary-terrorist-cum-businessman was indeed proving that he was "a brilliant Soviet diplomat and above all an intelligent human being," as Trotsky characterized him.

As soon as the English agreement was concluded Leslie Urquhart, former British consul in Batum and chairman of the Russo-Asiatic Company, left for Moscow to meet with Krassin. Despite the general lack of progress in their talks and an appalling rate of inflation which pegged the ruble to the English pound at 120,000 in Moscow and 400,000 in Transcaucasia, there was a sudden flurry of interest in Russian oil stocks on the London Exchange. Those shares had declined to levels even lower than the darkest days of 1918 but they were certainly as tempting to speculators as the certificates peddled in Paris cafes. The *Economist* issued a stern warning: "Purchasers of shares—albeit at what appear low prices—are, however, taking much for granted in the present position as regards information regarding the condition of the properties and other essential points."[6]

The "other essential points" were brutal: typhus, cholera, and malaria stalked the land; there was widespread famine, savagery, and even cannibalism in the countryside. Industry and transport were a shambles, the factories and refineries deserted, the oil wells filling with water. Towns were depopulated, Petersburg reduced to a mere third its former size. Millions had perished. Russia was in ruins. The oil industry was dead.

When Lenin squarely faced the coming catastrophe and forced through the NEP—something no one else could have done and his last sweeping reform before suffering a crippling stroke—he invited the world's capitalists to what they hoped would be a feast or funeral. Foreign firms were asked to bid for large concessions, to build huge factories in the manner of old. Echoes of Peter the Great, of nineteenth-century expansion, but this time no Russians were welcome; those who fled could return only as "technical advisers," not a tempting title for a Nobel or a Hagelin or any of the other exiled industrialists.

When the first concession was granted—it was awarded to Great Northern Telegraph—rumors of other contracts spread through the corridors of the western oil companies like oil spouting from a Balakhany gusher. "The Bolsheviks are giving the British exclusive rights in Baku! . . . All the properties are being returned! . . . Seventy percent is going to be exported! . . . Standard signed a secret deal in Moscow! . . . Shell is moving back to Grozny!" The Soviets undoubtedly encouraged the rumors just as they later learned how to exploit the fears and jealousies of one company or country by negotiating or even signing preliminary agreements with another. At the outset, however—at the time NEP was inaugurated—Soviet leaders led by Krassin probably hoped to have one large organization take over the oil industry on a concession basis, something similar to the 1925 agreement with the Harriman interests who moved into the Georgian manganese mines, settling with a couple of dozen former owners while reviving an industry which before the war was providing 52 percent of the world's manganese. Harriman's deal shattered the U. S. State Department's policy against American loans to the Soviet Union but typified the not unusual expedient of western economic priorities taking precedence over political considerations.

But when it came time for the Soviets to negotiate concessions for the oil industry the price of monopoly dictated by Deterding and others was too high and the Soviets were forced to fall back on their own resources. The western oil companies, still confident that the regime would collapse, waited in the wings for the final act of the tragedy to unfold. But they took turns interrupting the performance, declaiming on the importance of their own roles. Between the acts they gathered to publicly pledge mutual trust and understanding while privately rewriting their own scripts. The first road show was at San Remo, an April 1920 extension of the Versailles peace talks covering the petroleum spoils of war. The English and French came to an agreement on Asia Minor, Galicia, Rumania, and their respective colonies as well as the Russian Empire where "the two governments will give their joint support to their respective nationals in their joint efforts to obtain petroleum concessions and facilities to export and to arrange delivery of petroleum supplies."[7]

With Lenin's NEP and the Soviet hunger for massive infusions of outside capital and technical assistance those "petroleum concessions and facilities" appeared more tempting than ever before as reflected on the London market. Word spread of Deterding's discussions with the Soviets and news of Shell's preliminary agreement with Krassin leaked out; the cries of the oil companies for another conference reached near hysteria. None was called but the oil men of the world took advantage of the April 1922 gathering in Genoa to push their claims and to treat with the Soviet delegates: Krassin, Chicherin, and Litvinov. The subject of oil was not on the agenda and the word was never mentioned in the official communiqués, but oil is what the conference was all about. Over a hundred fifty companies had filed claims with the Soviets for confiscated oil properties and foreign offices throughout the west were besieged with demands for action at Genoa. The meeting had originally been called to improve general relations between the Soviets and the west, eager to meet at the conference table—except for the isolation-struck Americans who only sent an observer. The west wanted economic advantages and some political accommodation. The Soviets were eager—in fact, desperate—to prove that NEP Russia was filled with Communists but not Communism.

For the oil industrialists there was no progress at Genoa. As they eagerly sharpened their knives to divide the petroleum pie they could not resist the temptation to swing a blade at the back of a neighbor or the throat of a competitor. The conference ended in greater confusion than it had begun but for Standard a clear warning was sounded. The British, led by Lloyd George, secured Soviet agreement to define a former owner of Soviet property as someone holding that property prior to the 1918 nationalization. Standard, its sights set squarely on the Nobel assets and power, suddenly found itself at a great disadvantage in its continuing struggle with Deterding.

From Genoa the feuding parties moved to The Hague and when the conference there had similarly negative results Standard's Teagle and Asche, along with Riedemann from Germany, met with Deterding and Gustav in London to discuss joint action. Teagle drafted the "London Memo" summarizing their discussions and enunciating their common cause. It was not an agreement—no one signed it—but an understanding which would hopefully reduce if not eliminate the individual plea bargaining with the Soviet authorities. Typical of this era, there was no mention of Russia in the document. Nor was there any binding action forcing adherence to the main points of the memorandum: no independent negotiations, complete indemnification for confiscated properties or restitution with full compensation for damages, thirty-day notice for any deviation from the understanding.

A purposeful nobly-worded document but weak and toothless—it was a beginning, however, and Gustav ran with it to the other companies, urging compliance. Two months later he met in Paris with delegates from Standard and fourteen other companies to discuss specifics. The result was the Front Uni, an international consortium of the major oil companies. Gustav and most of the other delegates were certain that the Soviets would never be able to recover from their difficulties without major assistance from the western companies. Without that aid the oil industry would remain in ruins, in the state described by a recent visitor: "Misery has gripped the workers, transport is disorganized and what was once one of the great industrial communities of the world is now in chaos."[8]

Front Uni guaranteed that assistance to lead the Soviets out of that chaos would not be given; it also ensured that the Bolsheviks would be barred from selling their oil on the world market—without the foreign currency from oil sales they would be unable to purchase equipment and supplies essential for reconstruction. It was solemnly agreed by those at the conference that no Soviet oil would be purchased and that none of "the interested parties should prejudice directly or indirectly existing interests and vested rights of other owners dispossessed by the Soviet government."[9] Front Uni would thus hasten the day when the property would be returned to its owners; but to cover the possibility that such a day might never come there was also a proviso that 5 percent of each company's oil-sale profits would be placed in a fund for the compensation of those who had lost property.[10] No wonder Gustav threw himself so vigorously into the cause of Front Uni, serving as main negotiator between the different groups, acting as cheerleader to those flagging spirits who did not have as much to gain or otherwise lacked motivation and conviction of the worth and potential of the common goals.

It was all in vain. Front Uni was simply not a realistic, viable mechanism. The Soviets were too clever at the capitalists' own game. They continued to play one company against another, offering their oil well below world prices. In February 1923, less than six months after Front Uni was launched, Deterding contracted for the purchase of seventy thousand tons of Soviet kerosene with an option for an additional hundred thousand tons. His competitors were infuriated by the double-dealing but for Standard that anger was more feigned than real, a reaction to the realization that they had been outflanked. Standard's own man was then in Moscow trying to make his own deal with the Communists. Another American, one of the most flamboyant in a field that specialized in the breed, was en route to Moscow to seek his own fiefdom. He raised Nobel's and Standard's level of frustration to the breaking point.

Harry Sinclair of the Oklahoma fields, making the pilgrimage to the Bolsheviks, traveled in a manner befitting a Mantashev with an entourage that included a Roosevelt and a U.S. senator and needed two decks of an ocean liner, a $200,000 chartered train, a palace in Moscow. The Soviets were impressed but were not about to give away their entire industry as Sinclair—like Deterding before him—demanded. They did offer him all of Baku and the northern half of Sakhalin—an insult to the Japanese, who were furious, but fitting for Sinclair and the ill-fated Senator Fall: the area was a former tsarist penal colony. But Sinclair had to agree to invest $115 million and to arrange an American loan. After all, Sinclair was a good friend of President Harding and carried considerable weight in Washington. To the Soviets it seemed a reasonable *quid pro quo* but when Harding died and the news of the Teapot Dome scandals broke in the press Sinclair had to hurry home. As the tempest swirled in Washington the Soviets lost interest in his wheeling and dealing and Standard breathed a little easier—many oil men were convinced that Standard had leaked the initial information on Teapot Dome to the senate in order to catapult the troublesome Sinclair out of their Russian sphere of influence.

By that time Teagle and his board of directors were becoming increasingly pessimistic about their investment in Russia. Gustav continued to trumpet the cause of Front Uni, to argue that only by concerted action and constant boycott could the Bolsheviks be brought to heel. It was a losing battle. A year after Deterding opened the floodgates the French started buying Russian oil for their fleet; then came the English and the Italians. Fascist Italy became one of the most important purchasers and by 1927–1928, with 6 percent of world production, the Communists were selling 2.7 million tons to twenty-two different countries. Nearly a third of the total went to England and Italy.

The increased sales reflected not only the triumph of Krassin and the Communists but the victory of the industry itself over tremendous odds, the recovery from the abuses of war, revolution, and civil strife. There had been no major structural damage in the fields during this period, but when the Bolsheviks took over in 1920 what activity Lessner, Wannebo, Malm, and the officials of the other companies had been able to generate during all the troubles almost immediately terminated. From 1920 until 1923 drilling practically ceased and the wells filled with water. There was less than one percent of the drilling there had been when Russia led the world in production in 1900. The yield of the producing wells did, however, give the new masters a rate of production comparable to that before 1890. But only a fourth of the prewar work force of forty thousand remained and there were acute shortages of pipe, casing, spare parts of all kinds, and a total breakdown in the electrical supply system. When the Soviets organized their petroleum trust, Sayouzneft, they were faced with an industry dying from neglect but still crucial to their economic recovery.

The trust divided the production areas into three separate sectors: Embaneft for the Urals, Grosneft for the Grozny area, and Asneft for Baku where a former Nobel engineer A. P. Serebrowski was put in command. Recovery was negligible until NEP aid and investment started to arrive. Massive shipments of western—especially American—equipment streamed to the Soviet fields. General Electric, Westinghouse, and the Swedish ASEA repaired and replaced generator stations, and in a half-dozen years the fields were completely electrified—before the war only 20 percent benefited from electricity, primarily Nobel and Bnito. Other American advisors and technicians fanned out across the Soviet installations. American Barnsdall Corporation, connected to Sinclair and financed in part by Standard Oil of Indiana, arrived with a team of drillers and engineers to repair pumps and other gear and to instruct the Russians in the newest techniques of rotary drilling.[11] The Russian industry had come full circle in the fifty years since Ludwig Nobel brought over that team of American drillers. A later generation of Nobels had pioneered the use of the rotary drill and by 1928 the entire Soviet industry was converted from the old labor-intensive percussion method, a tenth as fast as rotary.

Barnsdall's contract called for the corporation to import $1.5 million of American machinery, to be paid $50,000 for each productive well, and to receive 15 to 20 percent of the oil produced over a fifteen-year period. But it remained only two years and when it departed the industry was well on its way to recovery. From the low point of 1921 production steadily increased until by 1927 it reached prewar levels. Exports doubled then redoubled as the revenues from oil sales abroad became the largest single Soviet source of foreign currency, accounting for more than 20 percent of the dollar value of all exports. Lenin's NEP and the shrewdness of his oil negotiators had dramatically revived the industry, keeping it firmly in Soviet hands.

The success sealed the fate of the Nobels. Each shipment of Soviet oil, each transfusion of foreign capital, settled the Swedes deeper into their involuntary exile. Faith and hope die hard in the breasts of those pulled from pedestals of power but the Nobels had to face the hard reality that they would never again rule their Russian empire.

CHAPTER SEVENTEEN

Exit Nobelevski

The realization that all was lost in Russia was of course as painful to Emanuel, who had long enjoyed the towering peaks of power and prosperity, as it was to Gustav, who had tasted a much smaller sampling of the joys and glories of being a Nobel in Russia and was still in training to take command of an empire. Each adjusted to its inevitability: Gustav by his continued activities to revive the spirit and substance of Front Uni while handling affairs in Egypt, France, Poland, Finland, arranging the annual meetings, preparing reports, working with SAIC and Standard; Emanuel with his slower, measured pace still serving as presiding officer, traveling frequently, busying himself with other investments and the plans of his brothers. In 1921 he moved back to Sweden, living rather modestly in three rooms on the fourth floor of a Stockholm apartment next to sister Marta and her family, taking the ferry a few minutes each day to cross the water to his office near the Grand Hotel. In 1923 he again became a Swedish citizen.

When the Nobels returned to their native land it was a time of serious depression across Europe, the continent's "darkest hour" according to the London *Economist*.[1] In Sweden as the brief postwar boom ended wages dropped between 15 and 20 percent, unemployment mounted, trading on the stock exchange hit historic lows, and the fires were banked in over 80 percent of the iron foundries. The Shipowners Association called it their worst year on record.[2] It was hardly a promising time for Emanuel and his clan to commence new careers in the mother country, to charter new corporations like Ludwig Alfred's Svenska Nobel-Diesel Corporation conceived originally as a means of keeping the Petersburg engineers together working as a team, perfecting the diesel engines they were developing, ready in an instant to return to Russia and their drawing boards in the Sampsonievsky factory.

Ludwig had been rallying the exiles around him ever since his 1917 arrival in Stockholm and when Emanuel, the banker Wallenberg, and other financiers put up the necessary cash in 1918 a small building was purchased in Nynäshamn about forty miles southeast of Stockholm and Nobel-Diesel established. By 1921 they had their engine perfected—1,600-horsepower, four-cylinder, two-stroke, directly reversible, and designed for installation in the larger ships. It had a measured fuel efficiency of over 80 percent, more than any other two-stroke engine.

In 1923 the engine was installed in a new *Zoroaster* built at Sweden's Gotaverken and commissioned by a Nobel-Hagelin concern called Rederi A. B. Zoroaster. The last ship ever built for a Nobel, it was the third of the fleet—which just five years earlier numbered three hundred fifteen—to honor the memory of the ancient fire-worshipers. It was taken around the world as a floating sales exhibit for the Nobel-Diesel engine and successful licensing arrangements were made with Mirrless, Bickerton and Day in England, Nydqvist and Holm in Trollhätten, Burgerhout's Machinenfabriek en Scheepswerf in Rotterdam, and the Niigata Takkosko firm in Tokyo. But these several contracts were not sufficiently rewarding to warrant further development in the rather cramped quarters at Nynäshamn and in 1925 the factory was closed down and future work on the engines left to the licensees. The *Zoroaster,* in a final twist of fate, was sold to the Soviets who put duplicates of the Nobel two-stroke engine in their new transport ships, the *Aleksey Rykoff* and *Yan Rudzutak* launched in 1925 and 1928. As ironic a conclusion to the Nobel pioneering work with diesels as the sale of Emanuel's sleek yacht *Gryadoustiy* to a Finn named Algot Niska who used all its speed to outrun the customs officials. Niska was a famous smuggler of alcohol.

After Nobel-Diesel closed its door Ludwig Alfred turned for a time to experimenting with production of stainless steel but was as unsuccessful in this venture as Emanuel was with his new enterprise, and equally ahead of the times. Emanuel was involved in secret communications. During his last year in Petersburg a German inventor had offered him the patent rights for an automatic coding device but the company engineer who analyzed the machine was not impressed and he rejected what would later become the official cipher unit of the German government. Emanuel had other things on his mind in 1917 than exploiting patent rights for secret writing machines. But when he returned to Sweden he had time to consider the potential and practicality of communicating secretly with any of the Nobel offices in his empire. After the agreement with Standard was signed the practicality became necessity. Communications between Petersburg and New York would have to be as secret as those between the capital and Baku. Nobel-Standard did not want Royal Dutch or the government reading over its corporate shoulder.

The postwar world was on the threshhold of a new age in communications and Emanuel—true to family tradition—was standing at the door, eager to explore and stimulate what was then the smallest stirring of a new industry. Both he and Hagelin invested $40,000 in the Swedish Cryptograph Corporation organized in 1916 with 85,000 Swedish crowns and superb official contacts for the purpose of "the development, promotion and sale of all types of ciphering mechanisms." The company was the brainchild of an extraordinary and rather eccentric inventor Arvid Gerhard Damm. His concepts were correct, but as his various mechanical marvels failed to function expenses mounted and the company was kept afloat only by investment of still more thousands from Emanuel. In the next few years that investment tripled but Damm's "Universal Cryptographer" failed to become the "cornerstone of the universal cipher system" as company brochures promised. Damm had similar difficulties developing his telegraph units although patents were granted in a dozen countries and several were sold to the Swedish Telegraph System, Marconi, and the French Telegraph Company. Central European licensing arrangements were completed with Telefunken and the Swedish press waxed enthusiastic about Damm's dramatic discoveries, boasting that still another Swedish inventor-entrepreneur was on his way to establishing a worldwide enterprise.

Damm never really produced a reliable unit. His shelves filled with prototypes as company coffers emptied. Emanuel's impatience increased along with his investment and Hagelin moved onto the board of directors in order to exercise greater control of expenditures. He cut losses his first year from over 100,000 crowns to a more manageable 20,000 but both he and Emanuel wanted still stricter supervision. They selected Hagelin's son Boris as overseer. Graduate of the Stockholm Institute of Technology—his required term of practical training had been taken at the Petersburg factory working on the submarine diesels—and trained at ASEA during the war as preparation to take over the project for expanded electrification of the Baku fields, Boris had just recently returned from the United States where he had been working in Standard's Central Engineering Department—also in preparation for his future responsibilities in Nobel Brothers. Assignment as auditor and overseer in the Cryptograph workshop was to be an interim commission but for Boris it turned into a lifetime career.

He turned with little enthusiasm to the new field, frustrated along with all the other Swedes of high and low position waiting in Stockholm for the return to Russia; but before long he learned that he had talents for mechanical as well as electrical engineering, that he could see clearly where those before could only envision blurs and hazy solutions. His genius for invention was matched by skills in selecting associates, contacting customers. In diplomacy and tact he was a gentleman of Emanuel's old school. He had his father's shrewd business sense and his inventions revolutionized the field of secret communications; the "Hagelin Cryptos" trademark spread to countless code rooms all around the globe. Boris Hagelin assumed his rightful place in that Swedish Inventors Hall of Fame reserved for the country's keenest contributors to the progress of mankind: the two Ericssons, De Laval, Winquist and his ball bearings, Dalen and acetylene gas, Lindqvist's Primus stove, Johansson and his precision instruments, the two Nobels—Ludwig and Alfred.[3]

It was another Nobel who made it all possible and Emanuel gloried in the achievement of the man he regarded as a son. Like his father, Emanuel was delighted by the happiness and success of those around him. He was an amiable extrovert, kind and compassionate. In the grand old days in Petersburg a dinner party at his table was never complete unless the ladies received some memento of the occasion, usually a creation from Fabergé executed in accordance with Emanuel's own suggestion. On one such evening the ladies were presented with icicles to commemorate the passing of the Russian winter—rock-crystal pendants with a mat surface clustered with small diamonds in a frost design.[4]

Sensitive to the needs of others, he found great pleasure in life—even in exile. He refused to be consumed by anger and bitterness; he rejected reliance on recriminations unlike a Deterding who never seemed to tire of describing Bolshevik duplicity, attacking their claim to achievements that were really the heritage and contribution of the dispossessed. "Nationalism," Emanuel used to say, "is a pretty word for the meanest of actions." But he refused to succumb to stronger language, not even when the Soviets implicated him in the Russian banknote trial of 1930. The two Georgians on the stand were accused of forging millions of dollars' worth of "chervonetz" and the Soviet propaganda machine cranked out charges linking Emanuel.

It was a harsh and false echo of those days in Kislovodsk when he had actually been responsible for creation of a new currency but the new sounds from the new Russia were all part of the Stalinist hard line, a preview of the later trials. Defendants were paraded before a gullible public and charged with being in contact with the so-called Torgprom Trade and Industrial Center headquartered in Paris with Nobel, Mantashev, and other exiled Russian capitalists running an extensive sabotage and espionage network supported and guided by the intelligence services of the imperialist nations. There were trumped-up particulars and trumped-up confessions as the defendants pleaded guilty to receiving 1.5 million rubles from the center which was organizing intervention from the west in order to bring down the Soviet government and reestablish capitalism. Tied to the center in this nefarious plot were all the old enemies of the Bolsheviks: the Cadets, the Socialist Revolutionaries, even the Mensheviks—all supposedly working with Nobel and Mantashev to exploit the struggle of the Trotskyites against the forces of unity and Russian centralism.[5]

Emanuel ignored the charges. His few remaining years were not going to be wasted worrying about such absurdities or falling victim to a paralyzing pestilence of hate and frustration. There were more important things to do. He was, after all, a Nobel, the ranking Nobel, heir and guardian of a great family and tradition. And at no time was that more important than during the annual Nobel Prize ceremonies. As the individual who had done so much to make those prizes possible, the one who had stood up to family, friends, and the crown to guarantee survival of the spirit of his uncle's will, who could more appropriately take the hand of the winner and with a sincere and royally hospitable manner extend congratulations and good wishes?

He enjoyed the occasion tremendously, the pomp and circumstance, the honor it did his uncle, the family, the nation, the memories it brought flooding into his own being. But then he enjoyed many things during his years of retirement, the many trips—especially to places he had never seen like India and Ceylon. There was the annual taking of the waters at Professor Dapper's Sanitorium in Bad Kissingen, long a mecca for the wealthy of Petersburg. His last trip was to Copenhagen in January 1932 to visit the tsar's sister, Grand Princess Olga Alexandrovna. Five weeks later he suffered a stroke and on 31 May he died.

The company lived on another quarter-century with Gustav and then Marta's son Nils Oleinikoff administering the last rites, settling with shareholders and lawyers around the world. The offices were moved to Stockholm to a small room on Värtavägen, an insignificant shadow of the former suites, smaller than any one of dozens of offices once scattered across the Nobel empire but adequate for discharge of the few remaining responsibilities. In 1937 Polish-Standard-Nobel was sold to Socony-Vacuum; the fields had proven unproductive averaging less than eight hundred barrels a day, seldom enough to supply the Libusza refinery. After initially heavy operating deficits there were a few years of profit but then decline and a return to the red. Nobel had neither the money nor the talent to put into the area and it was Standard who picked up the bills just as they did in Finland, but there the books continued to show a profit and there were fewer political problems. That too faded away and in 1959 Nobel Brothers Petroleum Company, in its eightieth year, was formally dissolved.

Nobelevski, the Nobel period of the world's oil history, was officially over. Marta was the last remaining direct link to Ludwig as those who remembered his visions and those who built so mightily on his foundations passed from the scene. Two of the most capable, Olsen and Hagelin, survived Emanuel. Olsen moved to Stockholm in the 1920s, selling his Oslo mansion to the Americans who installed their minister on the estate and made it their legation—it is still the residence of the ambassador. Hagelin remained active to the end which came in his ninety-fourth year—looking over his son's shoulder as Boris was carving out his own empire, serving on the board of the Finnish company until pensioned by Gustav whom he never really forgave for putting him to pasture at seventy-five; but always dreaming of the day when he might return to the land he loved, settling along the Volga, taking one last trip across the Caspian to the oil fields he had served so well.

He would have had little trouble recognizing the old Russia in the new. The more things change in Russia the more they remain the same: one empire replaces another, one autocrat another as the commissars cast out the barons, the Politburo replaces the princes, and the Chekists rise from the warm ashes of the Okhrana. But there were differences in degree. No tsar ever wielded the power of Stalin nor condemned to death so many millions of his subjects nor confiscated, terrorized, purged, and slaughtered with a maniacal efficiency defying description.

There were other differences. In a massive effort to obliterate a positive past from Soviet consciousness the names were changed. Petersburg, after a decade as Petrograd, became Leningrad as the Communists dedicated in the old capital no fewer than two hundred thirty-seven museums and plaques related in one way or another to Lenin and to the events of 1917. The Ludwig Nobel Factory, largest engine plant in the Soviet Union, became simply Russkiy Diesel and Emanuel's great clubhouse for the workers was converted into the Vyborg Palace of Culture. The office building of Nobel Brothers in town was taken over by the Coal Commissariat and a few blocks away the Kazan Cathedral where both Immanuel and Robert once had toiled was transformed into the Museum of Religious History and Atheism. Yuzovka, named for English industrialist Hughes, became Stalino and thirty years later Donetsk. Perm, near Ludwig's rifle plants, became Molotov. Nizhni-Novgorod was named for Gorky. Tsaritsyn, in a fitting trading of titles to honor the new Tsar of All the Russias, was rechristened Stalingrad—and years later rechristened Volgograd.

In the south at the headwaters of the Volga at Astrakhan the town of Kirovsky was put on the Soviet map as the old Azerbaidzhan capital of Elizavetpol was changed to Kirovabad; a smaller town to the west a few miles from Leninakan was renamed Kirovakan. In old Petersburg there suddenly appeared the Kirov Theatre, the Kirov Ballet, the Kirov Stadium. The Putilov plant became the Kirov Works. All this to honor the onetime secretary of the Azerbaidzhan Communist Party Sergei Kirov who, along with Mikoyan, was on the Red Army armored train that crushed the world's first Moslem republic. Kirov's assassination in Leningrad under circumstances still not thoroughly understood provided Stalin a martyr hero and a pretext for his purges of the 1930s.

In Baku, as all visual proof of the Nobel name was removed, Tagiev's textile factories overnight became the Lenin Mills and out where the benevolent Tatar's ships used to dock near the Nobel fleet there suddenly appeared the Paris Commune Shipyards. Villa Petrolea was taken over to accommodate the new commissars and the three-story house of cards built by Isa-Bey was turned into the oriental propaganda division of the Third International.

High on the hill overlooking Baku today, on top of the Friendship Restaurant in Kirov Park, there is an oversize statue of Kirov looking out over Kirovsky Boulevard, the Kirov State University, and the harbor Gorky once declared to be lovelier than the Bay of Naples. The site is ideal for a memorial to Ludwig Nobel, the man whose genius and perseverance meant so much to the city and the country. But in a land so insecure about its own past such a fitting remembrance would not serve Soviet history or Soviet hagiography. All traces of Nobel had to be removed, the pre-Soviet slate wiped clean.

Wrenched from a place in history, reduced to the category of nonperson, Immanuel, Ludwig, Robert, Emanuel, Hagelin, Olsen, and all those Nobelites who struggled and strained to build and create disappeared into the mist, their glory, power, and honor obliterated by those who seized what they had created, their acts long forgotten by those secure from such a miserable fate.

Notes

CHAPTER ONE

1. Frederick Wagner, *Submarine Fighter of the American Revolution: The Story of David Bushnell* (New York: Dodd, Mead, 1963), pp. 35–90.
2. Or perhaps because of Samuel Colt whose excessive secrecy about his submarine battery and mine-firing techniques alienated military professionals whose cooperation was a necessary prerequisite for War and Navy Department support—as convincingly demonstrated in Philip K. Lundeberg, *Samuel Colt's Submarine Battery: The Secret and the Enigma* (Washington, D. C.: Smithsonian Institution, 1974).
3. Bertil Almgren et al., *The Viking* (Stockholm: Tre Tryckare, 1972), pp. 25, 68–69, 132–40.
4. Werner Keller, *East Minus West = Zero: Russia's Debt to the Western World 1862–1962* (New York: Putnam's, 1962), pp. 87, 102.
5. *Kort Anvisning for Resande från Tyskland till Köpenhamn, Stockholm och Petersburg* [Short guide for travelers from Germany to Copenhagen, Stockholm and Petersburg], in Swedish and German (Stockholm, 1844), pp. 112–38.
6. The accounts of the meetings with Schilder and Jacobi and Immanuel's work with mines are based on the autobiographic manuscript written by Immanuel in his later years and now on deposit in the Nobel Archives at the Landsarkiv in Lund, Sweden (hereafter cited as Nobel Archives). Although brief and fragmented, the manuscript is a valuable source for Immanuel's years in Russia.
7. The timing of the grant to Nobel is interesting when placed in the perspective of the activities of a Russian naval mission in the United States between 1837 and 1841. While primarily interested in naval technology and steam navigation members of the mission did contact Samuel Colt, then working on underwater mines, inviting him to put his talents at the disposal of the tsar. Colt turned down the proposal, apparently in July 1841, and the mission sailed for home aboard the American-built steam frigate *Kamchatka* in September of that

year; Lundeberg, pp. 18, 74n. That September Immanuel was given official approval to proceed with mine development and in March 1842, well after the results of the mission and Colt's refusal of the offer had been reported and analyzed, Immanuel was awarded the 3,000 rubles.

8. The tale is also told that the reason for the bulge in the straight-line roadbed at Vishniy Volochek is because the tsar's pencil hit his finger holding down the ruler.

9. A. V. Yarotskiy, *Pavel L'vovich Schilling* (Moscow: Akademi Nauk SSSR, 1963), pp. 11–15, 22–34. A. A. Samarov and F. A. Petrov, ed., *Razvitie minnogo oruzhiya v russkom flote: Dokumenty* [Development of mine material in the Russian navy: documents] (Moscow: Voenno-morskoe izdatel'stvo voenno-morskoe ministerstva soiuza SSSR, 1951); ibid., pp. 1–105, is a valuable source for Russian pioneering breakthroughs in mine warfare although the role of Immanuel Nobel is greatly minimized.

10. In Soviet texts Moritz Hermann von Jacobi is transliterated to Boris Semenovich Yakobi and given the highest marks for his work with electricity and mines; e. g., A. V. Khramoi, *Ocherk istori razvitie avtomatiki v SSSR, Dooktyabri'ski period* [History of automation in Russia before 1917] (Moscow: Izdatel'stvo Akademi Nauk SSSR, 1956) pp. 30–31, 38, 80–83, 227.

11. The boys may have had the same Swedish tutor as the young Whistler; he was a teacher who instilled in the future painter a great admiration of Charles XII at the expense of Peter the Great. Roy McMullen, *Victorian Outsider: A Biography of J. A. M. Whistler* (New York: E. P. Dutton, 1973), pp. 25–26.

12. Although there is no supporting evidence it can safely be speculated that Alfred also met with another prominent Swedish-born and -educated engineer John William Nyström who published his *Treatise on Screw Propellers and Their Steam Engines* in Philadelphia in 1852.

13. Marta Nobel-Oleinikoff, *Ludvig Nobel och Hans Verk* [Ludwig Nobel and his work] (Stockholm: by the author, 1952), p. 57. This dismissal of Jacobi, described as "sluggish in thought and action," is as exaggerated and one-sided a characterization as the Soviet relegation of Immanuel to an insignificant footnote in the Soviet publication on naval mine developments, Samarov and Petrov, p. 104.

14. H. Noel Williams, *The Life and Letters of Admiral Sir Charles Napier* (London: Hutchinson, 1917), p. 260.

15. Major General Elers Napier, *The Life and Correspondence of Admiral Sir Charles Napier*, 2 vols. (London: Hurst, Blackett, 1862), 2:263–64, 329.

16. Henry Norton Sulivan, *Life and Letters of the Late Admiral Sir Bartholomew James Sulivan* (London: John Murray, 1896), pp. 288–308. One of the Nobel mines retrieved in the Baltic in 1855 and then decorated with the imperial eagle and labeled "Russian Infernal Machine" now hangs over the doorway between east wing galleries 20 and 21 in the British National Maritime Museum at Greenwich.

17. Henrik Schück and Ragnar Sohlman, *Nobel, Dynamite and Peace* (New York: Cosmopolitan, 1929), p. 41.

CHAPTER TWO

1. Ministry of Finance, *The Industries of Russia*, 5 vols. (St. Petersburg, 1893), 1:179–81.
2. Forrest A. Miller, *Dmitrii Miliutin and the Reform Era in Russia* (Nashville, Tennessee: Vanderbilt University, 1968), pp. 26–28, 53–57. A valuable source for this period drawing heavily on the seminal Russian studies by Moscow University Professor Pyotr Zaionchkovsky.
3. *The Mechanical Factory Ludwig Nobel 1862–1912* (St. Petersburg, 1912), pp. 10–15.
4. Joseph E. Smith, *Small Arms of the World*, 10th ed. (Harrisburg, Pennsylvania: Stackpole, 1973), pp. 48–59.
5. Keller, p. 47.
6. Smith, pp. 68–70.
7. Austrian War Department, *Die Wehrkraft Russlands* [The armed strength of Russia] (Vienna, 1871), p. 94. Ibid., pp. 112–13, reports that thirty thousand troops in the battalions of rifles were armed with Berdans, and in the infantry units four hundred thousand had the Krnka and two hundred nine thousand the Carle. Thirty thousand marines used the Barancov and local troops still had to rely on the muzzle-loading .59-caliber smooth-bore weapon.
8. Nobel-Oleinikoff, pp. 172–73.
9. Herbert Barry, *Russia in 1870* (London: Wyman and Sons, 1870), pp. 256–57.
10. Ministry of Finance, 1:146–52; Nobel-Oleinikoff, pp. 174–78.
11. Smith, p. 68.
12. John E. Parsons, *The First Winchester* (New York: Morrow, 1955), pp. 86–91.
13. But not quite forever for the Russians. Aleksandr Solzhenitsyn reported in *The Gulag Archipelago* (New York: Harper and Row, 1974), pp. 240–41, that in the earliest stages of World War II Stalin "threw the flower of Moscow's intelligentsia into the Vyazma meat grinder with Berdan single-loading rifles, vintage 1866, and only one for every five men at that. . . . What Lev Tolstoi is going to describe *that* Borodino for us?"
14. *Mechanical Factory Ludwig Nobel*, pp. 14–19.
15. Herbert Barry, *Ivan at Home* (London: Wyman and Sons, 1872), p. 177.
16. Anton Chekov, "The Three Sisters," in *The Works of Anton Chekov* (New York: Black's Reader Service, 1929), p. 549.

CHAPTER THREE

1. These studies, the patent applications, and clippings of Immanuel's newspaper articles are in Nobel Archives.
2. Erik Bergengren, *Alfred Nobel, the Man and His Work* (London: Thomas Nelson, 1962), p. 28.
3. Alexandre Dumas, *Adventures in Caucasia* (Philadelphia: Chilton, 1962), pp. 141, 150.

4. The standard work on the early history of petroleum exploitation is R. J. Forbes, *Studies in Early Petroleum History* (Leiden, Netherlands: E. J. Brill, 1958); ibid., pp. 155–62, covers Baku.

5. David M. Lang, *A Modern History of Soviet Georgia* (New York: Grove, 1962), p. 48.

6. Cassius M. Clay, *Memoirs*, 2 vols. (Cincinnati, Ohio: J. Fletcher Brennan, 1886), 1:539.

7. James Dodds Henry, *Baku, an Eventful History* (London: Archibald, Constable, 1905), pp. 30–35.

8. Ludwig Nobel to Alfred Nobel, 31 December 1874, Alfred Nobel Papers, Nobel Foundation, Stockholm (hereafter cited as Nobel Papers).

9. Henry, *Baku*, pp. 55–65; Sir Boverton-Redwood, *Petroleum*, 4th ed., 3 vols, (London: Charles Griffin, 1922), 1:6–7.

CHAPTER FOUR

1. Ludwig sent a copy of this memorandum to Alfred in early 1877 with a report on the status of dynamite sales in Russia. By the time Ludwig was ready to commence his project he had the results of an on-scene scientific survey of American techniques; the well-known Professor Dmitry Mendeleyev had been sent in 1876 by Alexander II on a study tour of the Pennsylvania fields and his report was published in Petersburg.

2. Charles Marvin, *Region of Eternal Fire* (London: W. H. Allen, 1884), pp. 283–307; Henry, *Baku*, pp. 72–74.

3. James Dodds Henry, *Thirty-five Years of Oil Transport* (London: Bradbury, Agnew, 1907), p. v.

4. In addition to Henry's essential study on oil transport (ibid.), the genesis and implementation of Ludwig's ideas are covered in some detail in idem, *Baku*, pp. 75–76; Marvin, *Eternal Fire*; Boverton-Redwood, 2:674, 679–88; Nobel-Oleinikoff, pp. 197–200. The contribution of Almqvist and the Lindholmen-Motala yards is covered in Harald Almqvist, "Sven Almqvist 1860–1910, Ett halvsekel i skeppsbyggeriets tjänst" [A half-century in the service of ship-building], *Unda Maris* (1967–1968), pp. 5–89. Typical of the later studies which totally ignore Ludwig's invention is Isaac F. Marcosson, *The Black Golconda* (New York: Harper, 1924); the historical review of tanker development does not even mention Nobel.

5. But when the Nobels and their company faded from the scene after 1917 the pioneering role of Ludwig and his contributions to marine transport also faded from memory. By the 1960s the Smithsonian Institution in its petroleum exhibit was identifying the *Glückauf* as "the first tanker" and in 1973 the EXXON Corporation's television commercial claimed that it was "the first vessel built expressly as an oil tanker." The prophet is not without honor in his own land, however: Sweden's Tingsryd Brewery recently included the *Zoroaster* in its beer-can series featuring full-color pictures of famous ships.

6. Henry, *Thirty-five Years*, p. vi.

CHAPTER FIVE

1. Ludwig's land and inland waterway distribution system was first reported in English by Charles Marvin in *Region of Eternal Fire* and several so-called shilling pamphlets: *Baku: The Petrolia of Europe, The Petroleum Industry of Southern Russia,* and *The Coming Deluge of Russian Petroleum and Its Bearing on British Trade.* Boverton-Redwood, 2:674–78, updated the information which is reported in greatest detail in the Nobel Brothers' twenty-five- and thirty-year anniversary reports, 1904 and 1909, the primary source used by Nobel-Oleinikoff, pp. 200–1, 250–62.

2. Schück and Sohlman, p. 80.

3. Nobel and Russian drilling techniques are discussed in Henry, *Baku,* pp. 117–18; Boverton-Redwood, 2:432–43; J. E. Brantly, *History of Oil Well Drilling* (Houston: Gulf, 1971), pp. 180–82. In none of these texts or in any other of the many sources consulted for this book has the author found any evidence to confirm or in any way support the September 1971 claim by Tass News Agency that the first Russian drilling for oil occurred in 1844, fifteen years before Colonel Drake's well in Titusville, Pennsylvania, traditionally accepted as the world's first drilled oil well.

4. The trailblazing Nobel refining methods and comparison with other distillation procedures are treated fully by Boverton-Redwood, 2:476-500, 538-59; Harold F. Williamson and Arnold R. Daum, *The American Petroleum Industry 1859–1899,* 2 vols. (Evanston, Illinois: Northwestern University, 1959), 1:263–73.

5. James Dodds Henry, *Oil, Fuel and the Empire* (London: Bradbury, Agnew, 1908), pp. 34–44.

6. Ibid., pp. 45–47; Boverton-Redwood, 3:949.

7. Marvin, *Eternal Fire,* p. 295.

CHAPTER SIX

1. Ludwig Nobel to Alfred Nobel, 7 May 1880; Nobel Papers.

2. Nobel-Oleinikoff, p. 289.

3. Fedor Dostoevsky, *Crime and Punishment* (Cleveland: Fine Editions, 1947), p. 424.

4. Schück and Sohlman, pp. 14–15.

5. Ivar Lagerwall, "Ludwig Nobel, En Industriens Storman" [Ludwig Nobel, giant of an industry], *Ord och Bild,* 20, no. 7 (July 1911):353–64.

6. Professor Troels Lund quoted in James W. B. Steveni, *Unknown Sweden* (London: Hurst and Blackett, 1925), p. 21.

7. Eli Heckscher, "Emanuel Nobel," in *Kungl. Svenska Vetenskapsakademiens Arsbok* (1933), pp. 295–304.

8. Nobel-Oleinikoff, p. 206.

9. Marvin, *Eternal Fire,* p. 250.

10. Schück and Sohlman, p. 8.

11. After that scare Ludwig was always quick to inform Alfred about any other fires that might affect the market or Alfred's nerves; for example, when fire damaged the Fedorov and Kolesnikov properties in Baku and some ships of the Drushina and Kawkas lines were burned Ludwig immediately notified his brother that no Nobel property was involved; Ludwig to Alfred, 29 May 1882; Nobel Papers.

12. Schück and Sohlman, p. 13.

13. Ludwig to Alfred, 30 December 1877; Nobel Papers.

14. The loan arrangements are detailed in a 9 March 1883 memorandum sent to Alfred along with production schedule estimates for the next nine months and delivery projections to the various storage depots for the next year; Nobel Papers. Years later Emanuel told Karl Hagelin that if the company had not managed to get this credit it would have been forced to declare bankruptcy. "It was much more difficult to get 40,000 rubles at that time than it was to get 4,000,000 ten years later"; Ida Bäckmann, *Från Filare till Storindustriell* [From apprentice to great industrialist] (Stockholm: Bonniers, 1935), p. 97.

15. The friend was Louis Berger, a successful businessman and veteran member of both the Prussian Landtag and German Reichstag. He had met Ludwig several years earlier when both were bidding for Russian military contracts, and he was the first to inform Ludwig about Bismarck's change in policy and his encouragement to the Berlin banks to extend credits to Russian firms; Lagerwall.

16. Ludwig to Alfred, 11 December 1883; Nobel Papers.

CHAPTER SEVEN

1. Frederic Morton, *The Rothschilds* (New York: Fawcett, 1963), pp. 97–104, 119, dramatically recounts the family's exploits in European railroading but one looks in vain for any discussion of Rothschild involvement in the Russian oil industry. However, for reasons similar to those which discouraged recording of the Nobel history, no one else has told the story either, although useful background information and insights are given in F. C. Gerretson, *History of the Royal Dutch*, 4 vols. (Leiden: E. J. Brill, 1953–1957), 1:213–14, 2:108–10; Oswald von Brackel and Joseph Leis, *Der Dressigjährig Petroleumkrieg* [The thirty years petroleum war] (Berlin: J. Guttentag, 1903), pp. 198–202; Robert Henriques, *Bearsted, a Biography of Marcus Samuel* (New York: Viking, 1960), pp. 72–75.

2. Charles Marvin, *The Coming Deluge of Russian Petroleum and Its Bearing on British Trade* (London: R. Anderson, 1887), p. 1.

3. Henry, *Thirty-five Years*, p. 199.

4. British Consul Patrick Stevens quoted in ibid., p. 120.

5. Henriques, pp. 75–76.

6. Beliamin was one of the original company shareholders (25,000 rubles) and

Lagerwall one of Ludwig's closest and most trusted advisors. He was also an enthusiastic admirer who capsulized his chief's genius in these words:

> To assemble the fruits of technical investigation, creative endeavor, and accumulated capital, convert them into the most impressive store of wealth, and then extravagantly to distribute these spiritual and material gifts, and to watch their transformation and their conversion into a gigantic structure under whose protection human life would become happier and civilization would be enriched, must have been a source of joy and satisfaction.

Schück and Sohlman, p. 72.

7. Fortunately for historians Lagerwall had been instructed by Ludwig to keep Alfred fully informed and his lengthy detailed letters were preserved in the Nobel Papers. Of particular importance are the 1884 letters of 14 and 17 June, 4 July, 7 August, 4 and 24 September, 4 October, and 11 November.
8. Summary report of negotiations with Rothschilds dated 2 June 1884, signed by Ludwig Nobel and forwarded to Alfred Nobel; Nobel Papers.
9. Lagerwall to Alfred Nobel, 24 September 1884; Nobel Papers.
10. Allan Nevins, *John D. Rockefeller, a Study in Power*, 2 vols. (New York: Scribner's, 1940), 2: 116.
11. Lagerwall to Alfred, 18 November 1885; Nobel papers.
12. Idem, 26 December 1885; Nobel Papers.
13. Reported to the author by Boris Hagelin who remembers with considerable pride the indignant manner in which his father and other Nobel officials reported and remembered such incidents.
14. The Bremen firm was Rassow Jung and Company, and a copy of the April 1886 report from Naftaport is in the Nobel Papers.

CHAPTER EIGHT

1. Gerretson, 2:108–9.
2. As indication of that concentration across the Caspian and along the Volga, of the 1885 production of 10.6 million poods from 1 January until the freezing over of the Volga in early November 9.9 million poods were shipped along the great river. Lagerwall to Alfred, 18 November 1885; Nobel Papers.
3. Henry, *Baku*, p. 2.
4. They couldn't even make it into the *Guinness Book of Records*. The "greatest gusher" is listed as the Spindletop well near Beaumont, Texas; during nine days in 1901 a "record" eight hundred thousand barrels spouted out of the ground; ibid., p. 289. Droozhba gave twice that amount before it was briefly capped, and four to five times as much as the Guinness record gusher if one includes all the oil that was allowed to run back into the ground.
5. Ralph Hewins, *Mr. Five Per Cent: The Biography of Calouste Gulbenkian* (New York: Rinehart, 1958), pp. 47–49. Gulbenkian's father and uncle had imported Russian oil to Turkey and Gulbenkian commenced his own fabulous

career in black gold by attaching himself like a limpet to Mantashev, serving as secretary, butler, and general factotum. See ibid., pp. 49–50; Glyn Roberts, *The Most Powerful Man in the World: The Life of Sir Henri Deterding* (New York: Covici-Friede, 1938), pp. 143–44.

6. Essad Bey, *Blood and Oil in the Orient* (New York: Simon and Schuster, 1932), p. 33.
7. Roberts, p. 49.
8. Marvin, *Eternal Fire*, p. 205.
9. Bäckmann, pp. 281–21.
10. Essad Bey, p. 42.
11. Nobel-Oleinikoff, pp. 321–22; Lagerwall to Alfred, 13 January 1886, and Ludwig's eight-page report of 20 January 1886 on company progress, the monetary difficulties, and rumors being circulated about the company's impending bankruptcy; Nobel Papers.
12. "Memorandum über das Consortium zur Erbauing einer Dynamitfabrik in Russland" [Memorandum on the consortium constructing a dynamite factory in Russia], 27 November 1878; Nobel Papers.
13. Nobel-Oleinikoff, pp. 318–19; Muriel E. Hidy and Ralph W. Hidy, *Pioneering in Big Business 1882–1911: The History of Standard Oil Company (New Jersey)*, 3 vols. (New York: Harper, 1956), 1:135–37.
14. Gulbenkian admired Lane as a "commerical genius and organizer"; Hewins, p. 55. For similar high praise see Gerretson, 2:103–4; Henriques, pp. 65–66.
15. Henriques, pp. 75–120, provides a detailed description of the Samuel debut in the petroleum trade but his unabashed eagerness to give his wife's grandfather Marcus Samuel all the credit must be balanced by less enthusiastic accounts such as those in Gerretson, 1:212–15.
16. Marvin, *Eternal Fire*, p. 303.

CHAPTER NINE

1. Bäckmann, p. 209.
2. Lagerwall to Alfred, 23 and 30 November 1889; Nobel Papers.
3. Ibid., 22 March 1889.
4. Ibid.
5. Ibid., 20 July 1889.
6. Standard was giving Nobel good reason to be concerned; it was winning back the north German market from Naftaport. The American consul in Stettin reported in 1891 that although Nobel kerosene was less expensive than the Standard brand it could not compete in popular esteem with "the superior American product." The consul advised Washington that only in south Germany, Turkey, and Austria-Hungary was the Russian product a competitive threat to the United States. Nevins, 2:122.
7. Hidy and Hidy, pp. 236–37, skip over these meetings; but see Henriques, pp. 126–55; Nobel-Oleinikoff, pp. 365–66; Brackel and Leis, pp. 199–202.

8. A copy of this document is in the Nobel Papers.
9. Theodore H. von Laue, *Sergei Witte and the Industrialization of Russia* (New York: Columbia University, 1963), pp. 108, 177; Richard Hare, *Portraits of Russian Personalities between Reform and Revolution* (London: Oxford University, 1949), pp. 301–29.
10. Sergei Meskhi quoted in Lang, p. 106.
11. For the production statistics in this and other chapters the following sources were of most value: the 1904 and 1909 Nobel Brothers' anniversary reports; Nobel-Oleinikoff, pp. 246–49; Hidy and Hidy, pp. 131–32, 237; Robert E. Ebel, *The Petroleum Industry of the Soviet Union* (New York: American Petroleum Institute, 1961), p. 74. A statistical gold mine on all aspects of the industry was written by the honorary secretary and executive council member of the Russo-British Chamber of Commerce in London, D. Ghambashidze, *The Caucasian Petroleum Industry and Its Importance for Eastern Europe and Asia* (London: Anglo-Georgian Society, 1918).

CHAPTER TEN

1. Karl Vasilievich Hagelin, *My Journey of Toil* (New York: Boris Hagelin, 1945), previously published in part in Swedish ten years earlier (Ida Bäckmann), provides detailed information on his career and valuable information on other company personalities and activities.
2. Although his memoirs were never published Hans Olsen also recorded details of his own career with Nobel Brothers and a copy of the eight-hundred-page typewritten manuscript in Norwegian, entitled "Livserindringer" [Life's recollections], is available at the Nobel Foundation. His commentary is essential to an understanding of the international negotiations during the Thirty Years' Petroleum War.
3. Henrik Schück and Ragnar Sohlman et al., *Nobel, the Man and His Prizes* (Norman, Oklahoma: University of Oklahoma, 1951), p. 50. Sohlman's essay in this book (ibid., pp. 47–84) is a detailed elaboration of the settlement of the estate and the squabbles within the family, briefly covered in idem, *Nobel, Dynamite and Peace*.
4. Nobel-Oleinikoff, p. 289.
5. Sohlman made a written memorandum of this interview as Emanuel reported it to him shortly after the audience with the king, and he included the account in his essays on Alfred's will: Schück and Sohlman, *Nobel, the Man and His Prizes*, pp. 76–77. In his later years, as reported to the author by Dr. Nils Oleinikoff, Emanuel related his memory of the interview as given in the last sentence.
6. Gerretson, 2:232–33.
7. Henriques, pp. 384–85.
8. Ibid., pp. 266–68, 370–79, 417–20.

CHAPTER ELEVEN

1. Schück and Sohlman, *Nobel, Dynamite and Peace*,
2. Bäckmann, p. 143.
3. Ibid., p. 163.
4. William B. Steveni, *Through Famine-Stricken Russia* (London: Sampson Low, Marston, 1892), p. 174.
5. Bäckmann, pp. 176–77.
6. Henry, *Baku*, p. viii; the society's activities are discussed in ibid., pp. 120–30. For supplementary information see Gerretson, 3:149–50.
7. Hewins, p. 25.
8. D. Ghambashidze, *Mineral Resources of Georgia and Caucasia* (London: Anglo-Georgian Society, 1919), p. 1.
9. Arthur Beeby-Thompson, *Oil Fields of Russia and the Russian Petroleum Industry* (London: Crosby, Lockwood, 1908), pp. 242–45.
10. Bäckmann, pp. 177–79.
11. Luigi Villari, *Fire and Sword in the Caucasus* (London: T. Fisher Unwin, 1906), p. 180.
12. Bäckmann, p. 162.

CHAPTER TWELVE

1. Sergei I. Witte, *The Memoirs of Count Witte* (Garden City, New York: Double-day, 1921), p. 198.
2. Bernard Pares, *A History of Russia* (New York: Knopf, 1950), pp. 398–99; Mikhail I. Tugan-Baronovsky, *Geschichte der russischen Fabriken* [History of Russian factories] (Berlin: E. Felber, 1900), pp. 401–19.
3. There is understandably an oversupply of material on the early revolutionary activities of Stalin but those with the most useful information on conditions and developments in Baku and the Caucasus are Isaac Deutscher, *Stalin, a Political Biography* (New York: Oxford University, 1949), pp. 27–108; Bertram Wolfe, *Three Who Made a Revolution*, 2 vols. (New York: Time, 1960), 2:117–60; Adam Ulam, *Stalin* (New York: Viking, 1973), pp. 16–113; Ronald G. Suny, "A Journeyman for the Revolution: Stalin and the Labour Movement in Baku June 1907–May 1908," *Soviet Studies* (January 1972), pp. 373–94; finally that psycho-history, "The Rendezvous of a Personality with the Public Political World," in Robert C. Tucker, *Stalin as Revolutionary 1879–1929* (New York: Norton, 1973), pp. 64 passim.
4. Wolfe, 1:329–30, 2:53–55, 68–70, 120–21, was not guilty of overlooking Krassin's career in and out of Baku; nor, more recently, was Ulam, pp. 59–63, who concludes with a speculative summary of Krassin's possible ties with Okhrana. The biography by his wife Lyubov Krassin, *Leonid Krassin, His Life and Work* (London: Skeffington & Son, 1929), sheds little light on the early clandestine career and is of most use for post-1917 activities.

5. Deutscher, p. 98.

6. Wolfe, 1:407–8.

7. Solomon M. Schwarz, *The Russian Revolution of 1905: The Workers' Movement and the Foundation of Bolshevism and Menshivism* (Chicago: University of Chicago, 1967), p. 130. Schwarz analyzes the myth and reality of Stalin's role in this strike; ibid., pp. 301–14.

8. Frederick A. Mackenzie, *From Tokyo to Tiflis* (London: Hurst and Blackett, 1905), p. 305.

9. Ibid., pp. 307–9.

10. Villari, pp. 180–200; Henry, *Baku*, pp. 149–63, 173–216; Maxwell S. Leigh, ed., *Memories of James Whishaw* (London: Methuen, 1935), pp. 130–32.

11. Deutscher, p. 66.

12. That cavalier analysis by Henry, *Baku*, p. ix, revealed as much sympathy for the striking workers as the judgment by Henry's countryman Arthur Beeby-Thompson, *Black Gold: The Story of an Oil Pioneer* (New York: Doubleday, 1961), pp. 72–73:

> Thus developed out of a few grievances, with a background of politics, a succession of incidents which almost ruined the great oil industry of Baku. So frivolous became the demands that the authorities were compelled to intervene with the aid of troops to check the wave of sabotage, incendiarism, intimidation and other forms of violence that originated from groundless charges made by agitators.

13. Deutscher, p. 100.

CHAPTER THIRTEEN

1. Schwarz, pp. 112–27, discusses the importance of this Shidlovskii commission and the meetings in Petersburg.

2. Nobel-Oleinikoff, p. 370.

3. Ibid., pp. 558–59.

4. Eugen Diesel, *Diesel, der Mensch, das Werk, das Schicksal* [Diesel, the man, the work, and fate] (Hamburg: Hanseatische, 1937), p. 287.

5. Ibid., pp. 287, 290.

6. Ibid., pp. 290–91, 395, 402; Eugen Diesel and Georg Strossner, *Kampf um eine Maschine* [The struggle for an engine] (Berlin: Schmidt, 1950), pp. 50, 144. Beer baron Adolphus Busch paid Diesel $250,000 for American rights and built the first diesel engine designed for industrial use, putting it into operation in September 1898 in his St. Louis brewery. By 1905 his American Diesel Engine Company had a line of three-cylinder models and Busch's diesels were installed in some hundred fifty plants and factories.

7. Photographs and descriptions of these units are found in the factory's fiftieth anniversary report in 1912, pp. 73–115, and in Anton Carlsund's manuscript copy, "Maskinfabriken Ludvig Nobel in St. Petersburg" [The Ludwig Nobel engine factory in St. Petersburg]; Dr. Nils Oleinikoff.

8. Bäckmann, pp. 228–32. The Soviets rediscovered Hagelin's all-Russian water route in July 1975 and Tass News Agency headlined the news as a "First Voyage from Helsinki to Iran via Internal Soviet Waterways." The ship, *Ladoga 9*, traveling through "the Saimaa Canal and along the Neva through lakes, canals and rivers, then along the Volga to the Caspian Sea," was expected to take three weeks to complete the trip.

9. Bäckmann, pp. 233–37; Ludwig Borggren, "Fran Vandal til Ymer" [From *Vandal* to *Ymer*] in *Daedalus, Tekniska Museets Arsbok* (1954), pp. 117–30. In 1905 the Swiss Sulzer firm unveiled a reversible diesel and five years later Vickers in England developed submarine engines. The first commercial marine diesel installations in the United States were completed in 1911 and the following year Burmeister and Wain in Denmark built an ocean-going ship powered by diesels.

10. See Borggren.

11. A. A. Kochetov, *Podvodnaya lodka 'Minoga,' opisanie* [Submarine Minoga, description] (St. Petersburg, 1910), provides a detailed analysis with sketches and photographs of the world's first diesel submarine.

12. *The Russian Imperial Navy 1914* (St. Petersburg: I. D. Sujtin, 1914), pp. 235–39, discussed Nobel contributions to modernization of the fleet; the Plotnikov-Noblessner story is covered in detail from a Soviet point of view in Grigoriy M. Trusov, *Podvodnye lodki v russkom i sovetskom flote* [Submarines in the Russian and Soviet fleets], 2nd rev. ed. (Leningrad: Gosudarstvennoe-soiuznoe izdatel'stvo sudostroitelnoe promyslennosti, 1963), pp. 216–49.

CHAPTER FOURTEEN

1. Henry, *Baku*, p. 155.

2. Gerretson, 3:155.

3. Boverton-Redwood, 2:442–43.

4. Henriques, pp. 489–95; Hidy and Hidy, pp. 554–65.

5. Olsen, p. 201.

6. Ibid., pp. 202–4; Gerretson, 4:119.

7. *Pall Mall Gazette Guide to the Oil Companies* (London: Pall Mall, 1910). The following year *Pall Mall* published *The Russian Oil Fields and Petroleum Industry* by John Mitzakis.

8. Gerretson, 4:136.

9. Cecil W. Ahlquist, "Das Ausländische Kapital in der Russischen Petroleumindustri" [Foreign capital in the Russian petroleum industry], a diploma thesis written in 1928 at Cologne University, is a forty-eight-page survey of the subject of foreign investment and of special value in its coverage of Nobel holdings—Ahlquist was Emanuel's nephew and had access to Nobel Company records; Dr. Nils Oleinikoff.

10. Bäckmann, p. 261.

CHAPTER FIFTEEN

1. Hagelin, pp. 358–63.
2. Branting, a future Nobel Peace Prize winner, had strongly opposed implementation of Alfred Nobel's will and establishment of the Nobel Foundation in a series of newspaper articles under the heading "Alfred Nobel's Will—Magnificent Intentions—Magnificent Blunder." Schück and Sohlman, *Nobel, the Man and His Prizes*, p. 53.
3. C. E. W. Petersson, *How to Do Business with Russia* (London: Pitman, 1917).
4. Louis de Robien, *Diary of a Diplomat in Russia 1917–1918* (New York: Praeger, 1970), pp. 132–222.
5. Bäckmann, p. 285.
6. Ibid., p. 282.
7. Hassan Arfa, *Under Five Shahs* (New York: Morrow, 1965), p. 83.
8. Bey, p. 129.
9. John Silverlight, *The Victor's Dilemma: Allied Intervention in the Russian Civil War 1917–1920* (New York: Weybright and Talley, 1970), pp. 95–99.
10. Arfa, p. 86.
11. Bey, pp. 279–86.

CHAPTER SIXTEEN

1. George S. Gibb and Evelyn H. Knowlton, *The Resurgent Years 1911–1927: The History of the Standard Oil Company (New Jersey)*, 3 vols. (New York: Harper, 1956), 2:323–28, 662.
2. Ibid., p. 643.
3. The lengthy negotiations are covered in ibid., pp. 328–35; Nobel-Oleinikoff, pp. 468–72.
4. Robert, p. 228; Frank C. Hanighen, *The Secret War (New York: John Day, 1934), pp. 91–92.*
5. Krassin, p. 18.
6. *The Economist, 21 May 1921.*
7. Francis Delaisi, *Oil: Its Influence on Politics* (London: Allen and Unwin, 1922), p. 92. This earliest of the many analyses of postwar petroleum politics—originally published in Paris in 1920—must be supplemented by readings in the later studies representing the spectrum of national and political interests. A juggling of the following provides a measure of understanding of at least the highlights from the many meetings and international conferences: Paul Apostol and Alexander Michelson, *La Lutte pour le pétrole et la Russie* (Paris: Payot, 1922); Robert Page Arnot, *The Politics of Oil* (London: Labour Publishing Company, 1924); E. H. Davenport and Sidney R. Cooke, *The Oil Trusts and Anglo-American Relations* (London: Macmillan, 1923); Ludwell Denny, *We Fight for Oil* (New York: Knopf, 1928); Pierre l'Espagnol

de la Tramerye, *The World Struggle* for Oil (New York: Knopf, 1924); Louis Fischer, *Oil Imperialism: The International Struggle for Petroleum* (London: Allen and Unwin, 1926); L. Vernon Gibbs, *Oil and Peace* (Los Angeles; Parker, Stone and Baird, 1929); Anton Mohr, The Oil War (New York: Harcourt Brace, 1926); Hanighen, *Secret War*; and the most balanced and thorough of the lot, Wilhelm Mautner, *Der Kampf um und gegen das Russische Erdöl* [The battle for and against Russial oil] (Vienna-Leipzig: Manzsche, 1929).

8. Marcosson, p. 121.

9. Hanighen, pp. 217–22.

10. The distributions from this fund and the termination of contributions into it were a great source of discord and dispute in the 1920s and 1930s and even today there is a lawyer working in Paris attempting to extricate some of the monies set aside in the 5-percent fund.

11. Antony C. Sutton, *Western Technology and Soviet Economic Development 1917–1930* (Stanford: Hoover Institution, 1968), pp. 16–44.

CHAPTER SEVENTEEN

1. *The Economist,* 17 September 1921.

2. Ibid., 24 September 1921.

3. The Hagelin success story, the Damm beginnings, and technical descriptions of Hagelin's cryptographers are found in David Kahn, *The Codebreakers* (New York: Macmillan, 1967), pp. 422–34.

4. Henry Charles Bainbridge, *Peter Carl Fabergé* (London: Spring, 1966), p. 58.

5. Louis Aragon, *A History of the USSR* (London: Weidenfeld and Nicolson, 1964), pp. 276–78; *Geschichte der Sowjetunion* [History of the Soviet Union] (Berlin: Rütten and Loening, 1961), pp. 406–7.

Bibliography

Almost all the public and private papers of the Russian Nobels were lost along with their other Russian property and the material patiently assembled during the 1930s in preparation for a major study was also destroyed when the family's country home in Kirjola was leveled during the Russo-Finnish war. But the historical record was saved from total oblivion.

The Landsarkiv in Lund, Sweden, has a valuable collection of documents dating from Immanuel's days in Russia including his many sketches and watercolors of buildings, mines, and machinery, a fragmented autobiographical essay, copies of contracts with the government, and numerous reports, maps, and sketches collected by Robert Nobel during his tenure as overseer in Baku.

Included in the papers of Alfred Nobel at the Stockholm Nobel Foundation are some five hundred letters from Ludwig and Emanuel, status reports by Lagerwall on international negotiations, company financial reports, statements on Alfred's investments in Nobel Brothers, a series of bank circulars floating company debentures, and the indispensable special company publications—richly illustrated oversize books printed to mark anniversaries: *The Mechanical Factory Ludwig Nobel 1862–1912, Nobel Brothers Petroleum Production Company 1879–1904*, and *Thirty Years of Activity of the Nobel Brothers Petroleum Production Company 1879–1909*, all in Russian.

The foundation also has a copy of a 1939 autobiography "Livserindringer" written by Hans Olsen but never published, and copies of two works privately printed: Marta Nobel-Oleinikoff, *Ludvig Nobel och Hans Verk* (Stockholm, 1952), and Karl Wilhelm Hagelin, *Moi trchdovoi pcht* [My journey of toil] (New York, 1945). Selections from Hagelin's book were published by Ida Bäckmann, *Fran Filare till Storindustriell* (Stockholm: Bonniers, 1935). Dr. Nils Oleinikoff has in his possession a copy of a 1932 short study by Anton Carlsund, "Maskinfabriken Ludvig Nobel in St. Petersburg," and a copy of Cecil Ahlquist's 1928 Cologne University thesis, "Das Ausländische Kapital in der Russischen Petroleumindustri," along with several reports and publications of Nobel Brothers.

Articles and Pamphlets

Almqvist, Harald, "Sven Almqvist 1860–1919, Ett halvsekel i skeppsbyg-geriets tjänst." *Unda Maris* (1967–1968), pp. 5–89.

Askew, William C. "Russian Military Strength on the Eve of the Franco-Prussian War." *The Slavonic and East European Review*, 30 (1951).

Austrian War Department. *Die Wehrkraft Russias*. Vienna, 1871.

Borggren, Ludwig. "Fran Vandal till Ymer." *Daedalus, Tekniska Museets Arsbok* (1954), pp. 117–30.

Ghambashidze, D. *The Caucasian Petroleum Industry and Its Importance for Eastern Europe and Asia*. London: Anglo-Georgian Society, 1918.

————. *Mineral Resources of Georgia and Caucasia*. London: Anglo-Georgian Society, 1919.

Heckscher, Eli. "Emanuel Nobel." *Kungl. Svenska Vetenskapsakademiens Årsbok* (1933), pp. 295–304.

Lagerwall, Ivar. "Ludvig Nobel, En Industriens Storman." *Ord och Bild*, 30, 7 (July 1911): 535–64.

Marvin, Charles. *Baku: The Petrolia of Europe*, London: R. Anderson, 1884.

————. *The Petroleum Industry of Southern Russia*. London: R. Anderson, 1884.

————. *The Moloch of Paraffin*. London: R. Anderson, 1886.

————. *The Coming Deluge of Russian Petroleum and Its Bearing on British Trade*. London: R. Anderson, 1887.

————. *The Coming Oil Age*. London: R. Anderson, 1889.

Mitzakis, John. *The Russian Oil Fields and Petroleum Industry*. London: Pall Mall, 1911.

Otis, A. "The Petroleum Industry of Russia." *Trade Information Bulletin 263*. U. S. Department of Commerce, 1924.

Pall Mall Gazette. *Guide to the Oil Companies*. London: Pall Mall, 1910.

Suny, Ronald G. "A Journeyman for the Revolution: Stalin and the Labour Movement in Baku June 1907–May 1908." *Soviet Studies* (January 1972), pp. 373–94.

Books

Abercromby, John. *A Trip Through the Eastern Caucasus*. London: Edward Stanford, 1889.

Almgren, Bertil et al. *The Viking*. Stockholm: Tre Tryckare, 1972.

American Petroleum Institute. *History of Petroleum Engineering*. Dallas: Boyd, 1961.

Andersson, Ingvar. *A History of Sweden*. New York: Praeger, 1956.

Apostol, Paul, and Michelson, Alexander. *La Lutte pour le pétrole et la Russie*. Paris: Payot, 1922.

Aragon, Louis. *A History of the USSR*. London: Weidenfeld and Nicolson, 1964.

Arfa, Hassan. *Under Five Shahs*. New York: Morrow, 1965.

Arnot, Robert Page. *The Politics of Oil*. London: Labour Publishing Company, 1924.

Austin, Paul B. *On Being Swedish*. London: Martin Secker and Warburg, 1968.

Baddeley, John P. *The Russian Conquest of the Caucasus*. London: Longmans, Green, 1908.

Bainbridge, Henry C. *Peter Carl Fabergé*. London: Spring Books, 1966.

Barry, Herbert. *Russia in 1870*. London: Wyman and Sons, 1870.

———. *Ivan at Home*. London: Wyman and Sons, 1872.

Beaton, Kendall. *Enterprise in Oil: A History of Shell in the U. S.* New York: Appleton-Century-Crofts, 1957.

Bechhofer, C. E. *In Deniken's Russia and the Caucasus 1919–1920*. London: W. Collins, 1921.

Beeby-Thompson, Arthur. *Oil Fields of Russia and the Russian Petroleum Industry*. London: Crosby, Lockwood, 1908.

———. *Oil Field Development and Petroleum Mining*. London: Crosby, Lockwood, 1910.

———. *Black Gold: The Story of an Oil Pioneer*. New York: Doubleday, 1961.

Bergengren, Erik. *Alfred Nobel, the Man and His Work*. London: Thomas Nelson, 1962.

Beria, Lavrenti. *On the History of the Bolshevik Organization in Transcaucasia*. Moscow: Foreign Languages Press, 1949.

Bey, Essad (Leon Noussimbaum). *Blood and Oil in the Orient*. New York: Simon and Schuster, 1932.

Bonn, C. R. H. *The Oil Tanker*. London: Association of Engineering and Shipbuilding Draughtsmen, 1922.

Boverton-Redwood, Sir. *Petroleum.* 2 vols.; rev. 1926. London: Charles Griffin, 1896.

Brackel, Oswald von, and Leis, Joseph. *Der Dreissigjährig Petroleumkrieg.* Berlin: J. Guttentag, 1903.

Brantly, J. E. *History of Oil Well Drilling.* Houston: Gulf, 1971.

Browne, G. F., ed. *Memorials of a Short Life: Biographic Sketch of W. F. A. Gaussen with Essays on Russian Life and Literature.* London: T. Fisher Unwin, 1895

Bryce, James. *Transcaucasia and Ararat: Being Notes on a Vacation Tour in the Autumn of 1876.* London: Macmillan, 1877.

Campbell, Robert W. *The Economics of Soviet Oil and Gas.* Baltimore: Johns Hopkins, 1968.

Charques, Richard D. *The Twilight of Imperial Russia.* Fair Lawn, New Jersey: Essential Books, 1959.

Cowie, J. S. *Mines, Minelayers and Minelaying.* London: Oxford University, 1949.

Curtis, William Eleroy, *Around the Black Sea.* London: Hodden and Stoughton, 1911.

Curtiss, John Shelton. *The Army of Nicholas I 1825–1855.* Durham, North Carolina: Duke University, 1965.

Davenport, E. H., and Cooke, Sidney R. *The Oil Trusts and Anglo-American Relations.* London: Macmillan, 1923.

Delaisi, Francis. *Oil: Its Influence on Politics.* London: Allen and Unwin, 1922.

Denny, Ludwell, *We Fight for Oil.* New York: Knopf, 1928.

Deterding, Henri. *An International Oilman (as Told to Stanley Naylor).* New York: Harper, 1934.

Deutscher, Isaac. *Stalin, a Political Biography.* New York: Oxford University, 1949.

Diesel, Eugen. *Diesel, der Mensch, das Werk, das Schicksal.* Hamburg: Hanseatische Verlagsanstalt, 1937.

———. and Strössner, Georg. *Kampf um eine Maschine, die Ersten Dieselmotoren in America.* Berlin: Schmidt, 1950.

Dillon, Emile J. *Russian Characteristics.* London: Chapman and Hall, 1892.

———. *Eclipse of Russia.* London: J. M. Dent, 1918.

Ebel, Robert. *The Petroleum Industry of the Soviet Union.* New York: American Petroleum Institute, 1961.

Ehrenkrook, Friedrich von. *Geschichte der Seeminen und Torpedoes.* Berlin: E. S. Mittler, 1878.

L'Espagnol de la Tramerye, Pierre. *The World Struggle for Oil.* New York: Knopf, 1924.

Farson, Negley. *The Lost World of the Caucasus.* New York: Doubleday, 1958.

Fischer, Louis. *Oil Imperialism: The International Struggle for Petroleum.* London: Allen and Unwin, 1926.

Forbes, Robert James. *Petroleum and Bitumen in Antiquity.* Leiden, Netherlands: E. J. Brill, 1936.

——. *Short History of the Art of Distillation.* Leiden, Netherlands: E. J. Brill, 1948.

——. *Studies in Early Petroleum History.* Leiden, Netherlands: E. J. Brill, 1958.

——, and O'Beirne, D. R. *The Technical Development of Royal Dutch–Shell 1890–1940.* Leiden, Netherlands: E. J. Brill, 1957.

French, F. J. F. *From Whitehall to the Caspian.* London: Odhams, 1921.

Gerretson, Frederick C. *History of the Royal Dutch.* 4 vols. Leiden, Netherlands: E. J. Brill, 1953-1957.

Geschichte der Sowjetunion. Berlin: Rütten and Loening, 1961.

Gibb, George S., and Knowlton, Evelyn H. *The Resurgent Years 1911–1927, the History of the Standard Oil Company (New Jersey).* Vol. 2. New York: Harper, 1956.

Gibbs, L. Vernon. *Oil and Peace.* Los Angeles, California: Parker, Stone and Baird, 1929.

Golovine, Ivan. *Russia Under the Autocrat Nicholas I.* New York: Praeger, 1970.

Graham, Stephen. *Changing Russia.* London: John Lane, 1915.

Greger, René. *The Russian Fleet 1914–1917.* London: Ian Allan, 1972.

Halasz, Nicholas. *Nobel: A Biography of Alfred Nobel.* New York: Orion, 1959.

Hanighen, Frank C. *The Secret War.* New York: John Day, 1934.

Hare, Richard. *Portraits of Russian Personalities Between Reform and Revolution.* London: Oxford University, 1949.

Hassman, Heinrich. *Oil in the Soviet Union: History, Geography, Problems.* Alfred M. Leeston, trans., with added information. Princeton, New Jersey: Princeton University, 1953.

Heckscher, Eli. *An Economic History of Sweden.* Cambridge, Massachusetts: Harvard University, 1954.

Henriques, Robert. *Bearsted, a Biography of Marcus Samuel.* New York: Viking, 1960.

Henry, James Dodds. *Baku, an Eventful History*. London: Archibald, Constable, 1905.

———. *Thirty-five Years of Oil Transport*. London: Bradbury, Agnew, 1907.

———. *Oil, Fuel and the Empire*. London: Bradbury, Agnew, 1908.

Hewins, Ralph. *Mr. Five Per Cent: The Biography of Calouste Gulbenkian*. New York: Rinehart, 1958.

Hidy, Muriel E., and Hidy, Ralph W. *Pioneering in Big Business 1882–1911, the History of Standard Oil Company (New Jersey)*. Vol. 1. New York: Harper, 1956.

Hindus, Maurice. *Mother Russia*. New York: Doubleday, 1943.

Hoover, Calvin B. *The Economic Life of Soviet Russia*. New York: Macmillan, 1931.

Hubback, John. *Russian Realities*. London: John Lane, 1915.

Kahn, David. *The Code Breakers*. New York: Macmillan, 1967.

Kazemzadeh, Firuz. *The Struggle for Transcaucasia 1917–1921*. New York: Philosophical Library, 1951.

Keller, Werner. *East Minus West = Zero: Russia's Debt to the Western World 1862–1962*. New York: Putnam's, 1962.

Kendrick, Thomas D. *A History of the Vikings*. New York: Scribner's, 1930.

Khramoi, A. V. *Ocherk istorii razvitie avtomatiki v SSSR, Docktyabr'skii period*. Moscow: Izdatel'stvo Akademi Nauk SSSR, 1956.

Kochetov, A. A. *Podvodnaya lodka 'Minoga' opisanie*. St. Petersburg, 1910.

Kohl, Johann G. *Russia and the Russians in 1842*. London: Chapman and Hall, 1842.

Kornilov, Aleksandr A. *Modern History of Russia*. 2 vols. New York: Knopf, 1943.

Krassin, Lyubov. *Leonid Krassin, His Life and Work*. London: Skeffington and Son, 1929.

Landor, Arnold H. S. *Across Coveted Lands*. New York: Scribner's, 1903.

Lang, David M. *A Modern History of Soviet Georgia*. New York: Grove, 1962.

Larson, Henrietta; Knowlton, Evelyn; and Popple, Charles. *New Horizons 1927–1950, the History of Standard Oil Company (New Jersey)*. Vol. 3. New York: Harper, 1971.

Leigh, Marxwell S., ed. *Memories of James Whishaw*. London: Methuen, 1935.

Lisle, B. Orchard. *Tanker Technique 1700–1936*. London: World Tankship Publications, 1936.

Little, George H. *The Marine Transport of Petroleum*. London: E. and F. N. Spon, 1891.

Lodwick, John, and Young, D. H. *Gulbenkian.* New York: Doubleday, 1958.

Longhurst, Henry. *Adventure in Oil: The History of British Petroleum.* London: Sidgwick and Jackson, 1959.

Lundeberg, Philip K. *Samuel Colt's Submarine Battery: The Secret and the Enigma.* Washington, D.C.: Smithsonian Institution, 1974.

Mackenzie, Frederick A. *From Tokyo to Tiflis.* London: Hurst and Blackett, 1905.

Marcosson, Isaac .F. *The Black Golconda.* New York: Harper, 1924.

Marvin, Charles. *Region of Eternal Fire.* London: W. H. Allen, 1884.

Mautner, Wilhelm. *Der Kampf um und gegen das Russische Erdöl.* Vienna-Leipzig: Manzsche, 1929.

Mavor, James. *An Economic History of Russia.* 2 vols. London: J. M. Dent, 1914.

McMullen, Roy. *Victorian Outsider: A Biography of J. A. M. Whistler.* New York: E. P. Dutton, 1973.

Miller, Forrestt A. *Dmitrii Miliutin and the Reform Era in Russia.* Nashville, Tennessee: Vanderbilt University, 1968.

Miller, Margaret S. *The Economic Development of Russia 1905–1914.* London: P. S. King and Son, 1926.

Miller, Wright. *Leningrad.* New York: A. S. Barnes, 1970.

Ministry of Finance. *The Industries of Russia.* 5 vols. D. I. Mendeleev, ed., Russian edition; John Martin Crawford, ed., English edition. St. Petersburg, 1893.

Mohr, Anton. *The Oil War.* New York: Harcourt Brace, 1926.

Morell, R. W. *Oil Tankers.* New York: Simmons-Boardman, 1931.

Morton, Frederic. *The Rothschilds.* New York: Atheneum, 1962.

Mosley, Leonard. *Power Play: Oil in the Middle East.* New York: Random House, 1973.

Mosse, W. E. *Alexander II and the Modernization of Russia.* New York: Collier, 1962.

Muravyov, Ivan, et al. *Development and Exploitation of Oil and Gas Fields.* Moscow: Peace Publishers, n.d.

Napier, Elers. *The Life and Correspondence of Admiral Sir Charles Napier.* 2 vols. London: Hurst, Blackett, 1862.

Nash, Gerald. *United States Oil Policy 1899–1964.* Pittsburgh, Pennsylvania: University of Pittsburgh, 1968.

Nevins, Allan. *John D. Rockefeller, a Study in Power.* 2 vols. New York: Scribner's, 1940.

Oakley, Stewart. *A Short History of Sweden.* New York: Praeger, 1966.

Pares, Bernard. *A History of Russia.* New York: Knopf, 1950.

Parsons, John E. *The First Winchester.* New York: Morrow, 1955.

Petersson, C. E. W. *How to Do Business with Russia.* London: Pitman, 1917.

Pogue, J. E. *Economics of Petroleum.* New York: J. Wiley, 1921.

Post, Laurens van der. *A View of All the Russias.* New York: Morrow, 1964.

Ransome, Arthur. *Russia in 1919.* New York: Huebsch, 1919.

Roberts, Glyn. *The Most Powerful Man in the World: The Life of Sir Henri Deterding.* New York: Covici-Friede, 1938.

Robien, Louis de. *The Diary of a Diplomat in Russia 1917–1918.* New York: Praeger, 1970.

Ross, Edward Alsworth. *Russia in Upheaval.* New York: Century, 1919.

Russian Imperial Navy 1914. St. Petersburg: I. D. Sujtin, 1914.

Samarov, A. A., and Petrov, F. A., eds. *Razvitie minnogo oruzhiya v russkom flote: Dokumenty.* Moscow: Voenno-morskoe izdatel'stvo voenno-morskoe ministerstva soiuza SSSR, 1951.

Schück, Henrik, and Sohlman, Ragnar. *Nobel, Dynamite and Peace.* New York: Cosmopolitan, 1929.

———, et al. *Nobel, the Man and His Prizes.* Norman, Oklahoma: University of Oklahoma, 1951.

Schwarz, Solomon M. *The Russian Revolution of 1905: The Workers' Movement and the Formation of Bolshevism and Menshevism.* Chicago: University of Chicago (Hoover Institution Press), 1967.

Scott, George W. *The Swedes: A Jigsaw Puzzle.* London: Sidgwick and Jackson, 1967.

Seton-Watson, Hugh. *The Decline of Imperial Russia 1855–1914.* New York: Praeger, 1956.

Silverlight, John. *The Victor's Dilemma.* New York: Weybright and Talley, 1970.

Sleeman, Charles William. *Torpedoes and Torpedo Warfare.* Portsmouth, England: Griffin, 1881.

Smith, Edward Ellis. *The Young Stalin.* New York: Farrar, Strauss and Giroux, 1967.

Smith, Joseph E. *Small Arms of the World.* 10th ed. Harrisburg, Pennsylvania: Stackpole, 1973.

Snodgrass, J. H. *Russia: A Handbook on Commercial and Industrial Conditions.* Washington, D.C.: U.S. Department of Commerce, 1913.

Steveni, James W. B. *Unknown Sweden.* London: Hurst and Blackett, 1925.

Steveni, William B. *Through Famine-Stricken Russia*. London: Sampson Low, Marston, 1892.

Stewart, John Massey. *Across the Russias*. Chicago: Rand McNally, 1969.

Sulivan, Henry Norton. *Life and Letters of the Late Admiral Sir Bartholomew James Sulivan*. London: John Murray, 1896.

Suny, Ronald G. *The Baku Commune 1917–1918*. Princeton, New Jersey: Princeton University, 1972.

Sutton, Antony C. *Western Technology and Soviet Economic Development 1917–1930*. Stanford, California: Hoover Institution, 1968.

Tarbell, Ida. *The History of the Standard Oil Company*. New York: Macmillan, 1904.

Tjerneld, Staffan. *Nobel*. Stockholm: Bonniers, 1972.

Törnudd, Gustav. *I Oljans och Vindarnas Land*. Tammerfors: Süderstrüms, 1956.

Trusov, Grigoriy M. *Podvodnye lodki v russkom i sovetskom flote*. Leningrad: Gosudarstvennoe-soiuznoe izdatel'stvo sudostroitelnoe promyslennosti, 1963. 2nd rev. ed.

Truth's Investigator. *The Great Oil Octopus*. London: T. Fisher Unwin, 1911.

Tucker, Robert C. *Stalin as Revolutionary 1879–1929*. New York: Norton, 1973.

Tugan-Baronovsky, Mikhail I. *Geschichte der russischen Fabriken*. Berlin: E. Felber, 1900.

Tugendhat, Christopher. *Oil: The Biggest Business*. New York: Putnam, 1968.

Ulam, Adam. *Stalin*. New York: Viking, 1973.

Varney, John. *Sketches of Soviet Russia*. New York: Brown, 1920.

Villari, Luigi. *Fire and Sword in the Caucasus*. London: T. Fisher Unwin, 1906.

Von Laue, Theodore H. *Sergei Witte and the Industrialization of Russia*. New York: Columbia University, 1963.

Wagner, Frederick. *Submarine Fighter of the American Revolution: The Story of David Bushnell*. New York: Dodd, Mead, 1963.

Wallace, Sir Donald MacKenzie. *Russia on the Eve of War and Revolution*. Cyril E. Black, ed. New York: Random House, 1961.

Williamson, Harold F., and Daum, Arnold R. *The American Petroleum Industry 1859–1899*. 2 vols. Evanston, Illinois: Northwestern University, 1959.

Wilton, Robert. *Russia's Agony*. London: Edward Arnold, 1918.

Witte, Sergei I. *The Memoirs of Count Witte*. Abraham Yarmolinsky, trans. and ed. Garden City, New York: Doubleday, 1921.

Wolfe, Bertram. *Three Who Made a Revolution*. 2 vols. New York: Time, 1960.

Yarotskiy, A. V. *Pavel L'vovich Schilling*. Moscow: Akademi Nauk SSSR, 1963.

Zieber, Paul. *Die Sowjetische Erdölwirtschaft*. Hamburg: Gram, Gruyter, 1962.

Index

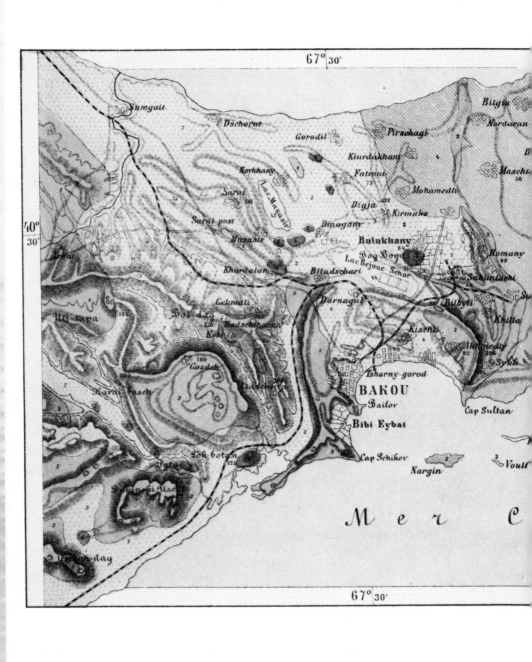